AF657490

PLANT ECOPHYSIOLOGY
A GLOSSARY

PLANT ECOPHYSIOLOGY
A GLOSSARY

NAFEES A. KHAN

Professor
Department of Botany
Aligarh Muslim University
Aligarh

R.M. AGRAWAL

Professor
School of Studies in Botany
Jiwaji University
Gwalior

I.K. International Publishing House Pvt. Ltd.

New Delhi • Bangalore

Published by
I.K. International Publishing House Pvt. Ltd.
S-25, Green Park Extension
Uphaar Cinema Market
New Delhi–110 016 (India)
E-mail: info@ikinternational.com
Website: www.ikbooks.com

ISBN 978-93-81141-15-1

Published by Krishan Makhijani for I.K. International Publishing House Pvt. Ltd., S-25, Green Park Extension, Uphaar Cinema Market, New Delhi–110 016 and Printed by Rekha Printers Pvt. Ltd., Okhla Industrial Area, Phase II, New Delhi–110 020.

Preface

Many standard textbooks are available dealing with the different aspects of plant physiology and ecophysiology which, however, are very large and exhaustive. *Plant Ecophysiology: A Glossary* will allow the students with an instant access to important terminology. For many topics and aspects related to plant ecophysiology, the students may find the book as a 'ready reckoner'/a useful revision aid and a stimulus for further study. The book is a comprehensive and self-contained work. The book has been prepared with an objective to be of use both for a layman and to a professional, however, in no way act as a replacement for specialist treatises. Authors shall welcome constructive criticism and shall put an effort to overcome the lapses and omissions in future.

We would like to thank our families for forbearance during the preparation of the book and our students, interactions with whom for around three decades have helped us understand the subject.

AUTHORS

A

Abscisic Acid

It is a sesquiterpenoid plant growth substance found in vascular plants. It is a plant hormone that allows a plant to adjust to the outside stress that it may encounter. It mainly acts to inhibit growth, promote dormancy and helps the plant to tolerate stressful conditions. It is so named because it causes the abscission of leaves from deciduous trees in the fall. It is also called dormin as it causes bud dormancy. In liverworts lunularic acid, a compound similar to ABA seems to play physiological role similar to that of ABA in higher plants.

Abscission

It is the shedding of a plant part. It refers to intentional dropping of one or more plant parts, such as leaf, fruit, flower or seed. Abscission occurs after the formation of an abscission zone, a thin layer of cells. The abscission layer becomes weak and breaks by converting pectic acid to pectin. Abscission process primarily seems to be regulated by ethylene and auxins suppress the ethylene effect. Nevertheless, greater auxin concentration may stimulate ethylene production.

Absolute Growth Rate

It is the dry matter accumulation (or energy content) per plant in unit time interval.

Absolute Temperature

It is the temperature measured in Kelvin from absolute zero. It is calculated on an absolute scale. Absolute temperature scale includes Kelvin and Rankine. The Kelvin scale is also called Absolute temperature and Kelvin is the SI unit for temperature. In this temperature scale 0°C = 273.15 K.

Absorption Spectrum

This is the absorption of different wavelengths by a molecule/substance. The intensity of the radiant energy at each wavelength gives the measure of the amount of energy at that wavelength. The absorption of radiation is measured with spectrophotometer when light passes through a selected absorbing substance.

Accessory Pigments

These are the pigments associated with both antenna and reaction centre pigment proteins and help in the transfer of absorbed energy to reaction centre during photosynthesis.

Acclimation

Acclimation or acclimatization is the adjustment to adverse condition. Plants that are continuously exposed to adverse conditions become tolerant compared to plants of same species that are not subjected to such adverse conditions. Acclimation of plant is important in agriculture. Generally, acclimation is distinguished from adaptation, which is used to indicate genetically determined level of tolerance.

Acid Growth Hypothesis

It is used to explain the rapid growth effects caused by auxin. The hypothesis states that auxin action causes H^+ ions accumulation, lowers the wall pH causing wall loosening and subsequently cell elongation and growth occurs. Generally cell walls extend much faster at acidic pH. At times auxin-stimulated cell elongation is attributed to increased proton extrusion and lowering of pH. The proteins named expansions have been implicated in pH-dependent extension of cell walls.

Acid Rain

It refers to the rain that contains oxides of nitrogen and sulphur. As a result of pollution, the oxides of nitrogen and sulphur are transformed into moist secondary pollutants such as sulphuric acid, ammonium nitrate and nitric acid that fall as rain, snow or fog. This rain is called acid rain. The transformation of these pollutants occurs because they can be transported to long distances and so can occur at places far away from the site of origin. Acid rain is harmful as it affects life on both land and water.

Acid Soil Complex

Acid rains contribute acid to soil. In soils which lack calcium carbonate plants are more adversely affected by these acids. Aluminium and/or manganese toxicity resulted under calcium deficiency conditions can be prevented by liming.

Acid Value

Number of halogens absorbed by 100 g of oil is referred to as acid value.

Acidic Soil

Soils that have pH less than 7 are said to be acidic soils. In such a soil aluminum and manganese can become so soluble so as to reach toxic level. Highly acidic soil loses cations rapidly and so becomes relatively poor. Soil can be acidic depending on the nature of original rock but rainfall is a determining factor in describing the nature of soil.

Acidophiles

The organisms which occur in highly acidic environments are acidophiles, e. g., *Thiobacillus thiooxidans.*

Action Spectrum

This is the relationship between the rate of light-dependent biological process and the wavelength of light. Generally it is in the form of a graph. The existence of the two photosystems in oxygenic photosynthesis is explained using action spectra.

Active Transport

It is a biological mechanism for the transport of molecules across a membrane against the concentration gradient. Active transport needs energy for transport of molecules either in the form of ATP or an electrochemical gradient. It is referred to as primary active transport if the same membrane protein is considered to break down the ATP using the energy so released in the transport of a solute across a cell membrane. In secondary transport solute movement is driven by the proton gradient resulted due to primary active transport ATPase.

Adaptation

Any morphological, physiological or behavioural characteristics which are considered to be evolutionary improvements that enhance the environmental fitness of organisms are considered adaptation. Genotypically determined characters for stress resistance such as sunken stomata, deep roots etc. determine whether the plants are stressed or not.

Adenylate Cyclase

The enzyme catalyzing formation of cAMP from ATP is adenylate cyclase.

Adequate Zone

Maximum growth at critical nutrient concentration is described as adequate zone. The growth is not increased even with the increase in nutrient concentration. Any further increase in nutrient concentration is only stored in vacuole. The storage is wider for macronutrients whereas narrow for micronutrients.

Adhesion

It is the attraction between unlike molecules.

Adhesive Forces of Water

It is the hydration force between water molecule and cell wall caused by hydrogen bonding. It is the force that plays an important role in the ascent of sap.

Adiabatic

The change in volume of air under constant temperature is called adiabatic change. Adiabatic expansion results in cooling of air at higher altitude because pressure decreases as we go up so air expands. If this expansion is adiabatic it will have less heat per volume and the air therefore will be cool.

Adiabatic Gradient

It refers to the gradient of adiabatic expansion or contraction. It is because of this gradient that as we go up pressure decreases and air expands at constant temperature making it cool.

Adsorption

It is a surface phenomenon in which water molecules are adhered to the surface of solid or rather any particle (solid, liquid or gas) is attracted to the surface of that particle. The attraction develops because of opposite charges on the two particles. Adsorption increases the surface area.

Aerobes

The organisms capable of growing in the presence of free oxygen.

Aerobic

It is the condition characterized by the presence of oxygen.

Aerobic Respiration

A biological process by which reduced organic compounds are oxidized is aerobic respiration. The energy released in the process can be utilized in the formation of ATP, which can be used for different processes involved in the maintenance of development.

Aflatoxins

Aflatoxins are secondary metabolites produced by fungus such as *Aspergillus flavus.*

Agronomy

It is the science that deals with field crops like cereals, forage, root crops and others.

Air Layering

The method of vegetative propagation in which aerial shoot is induced to form roots while still attached to the plant is air layering. After rooting it can be potted.

Albino

The condition of lacking chlorophyll or normal plant pigments.

Aleurone Layer

It is present around the endosperm of seeds and is a thin layer two- to four-cell thick. When the seeds absorbs water they germinate and after germination the aleurone cells provide hydrolytic enzymes that digest the starch, protein, phytin, RNA and other cell wall material present in the endosperm cells.

Alkaloids

These are compounds that contain nitrogen in a heterocyclic ring of variable structure. As nitrogen has a lone pair of electron, it can accept hydrogen ion and, therefore, alkaloids acts as base. They are aromatic, only slightly soluble in water and are mostly crystalline in nature. The first alkaloid isolated was morphine from *Papaver somniferum*. Mostly alkaloid synthesis in plants occurs in shoots but nicotine is produced only in the roots of tobacco. Alkaloids have medicinal value but their function in the plant that produces them is not clear. However, in some cases they protect the plant from grazing animals who avoid eating these plants because of its alkaloid content.

Alkalophilic

The organisms which show optimal growth in alkaline environments between pH 10 to 12 are alkalophilic.

Allelochemicals

It refers to chemicals that are secreted by plants. They allow interaction between organisms of different species, which may be positive or negative and acts at both physiological and behavioral level.

Allelopathics

The substances produced by one plant and harming another in allelopathy are called allelopathics.

Allelopathy

It is the phenomenon in which one plant harms the other plant growing nearby. This is the harmful influence exerted by a plant on the neighboring ones by the production of chemicals known as allelochemicals (released from plants or their residue). The allelopathic effects can be positive (inhibitory) or at times negative (stimulating) as well. Such chemicals are generally secondary metabolites, e.g., phenols.

Allosteric Compounds

The molecules reacting with enzymes in a way to inhibit or promote the enzyme activity are known as allosteric compounds. They do not react with the active site but with other part of enzyme, affecting the configuration of the active site.

Allosteric Effectors

These are small molecules that bind at site other than the active site called the allosteric site. Allosteric effectors enhance the catalytic activity of an enzyme.

Allosteric Enzyme

These are enzymes in which molecules bind to allosteric sites (sites different from the active site) and cause activation or inhibition of enzyme activity. The activity of these enzymes is modulated by two sites:

1) Active or catalytic site
2) Allosteric site

Allosteric Sites

These are sites in the enzyme other than active sites where combination with small molecules occurs. The enzyme binds with small molecules (allosteric effecter or repressor) at this site. Such a binding may be favorable in case the substrate is an allosteric effector and unfavorable in case of repressor. Binding at this site affects the catalytic activity of an enzyme.

Ambient CO_2 Concentration

It is the CO_2 concentration in the external air surrounding the canopy.

Amino Acid

These are amphoteric in nature having basic NH_2 (amino) and an acidic COOH (carboxyl) group. The two groups with an 'R' group that contain remainder molecule of amino acid and this 'R' differentiates one amino acid from other. Amino acids are linked by peptide bond in protein. The amino acid may be aliphatic (straight chain) or aromatic (ring structure). They may be acidic having free carboxyl group (aspartic and glutamic) or with basic amino group like arginine and lysine or sulfur-containing like methionine and cysteine.

Amino Group

These are NH_2 containing groups present in amino acids and impart basic property to the amino acid if they are present in the 'R' group, i. e., besides the –COOH group; $-NH_2$ group in amino acid. Amino groups are involved in amide formation, when the 'R' portion of an amino acid has an amino group linked to carbonyl carbon. These amino groups are involved in protein synthesis because the peptide bond is formed when the carbonyl group of one amino acid joins the amino group of another amino acid.

Ammonification

It is the process in which organic nitrogen is converted into NH_4^+ (ammonium ions) by microorganisms like bacteria and even some fungi and actinomycetes. The bacteria causing ammonification are known as ammonifying bacteria. The organic nitrogen present in animal excreta or plant debris is decomposed to release ammonia that is converted to NH_4^+ by bacteria.

Amylose

It is a polysaccharide consisting of D-glucose units joined to each other by α (1,4) glucosidic linkage. It gives blue color with iodine.

Anabolism

It is an energy requiring process in which large molecules are synthesized from small molecules, e.g., the process of photosynthesis is anabolic where small molecules like CO_2 and H_2O form glucose utilizing light from sun as a source of energy.

Anaerobic

Under conditions of oxygen deficiency, organisms accumulate NADH and pyruvate and carry out anaerobic respiration resulting in the formation of ethanol or lactic acid. An organism which can grow in the absence of oxygen is called an anaerobe, and such a phenomenon as 'anaerobiosis'. Anaerobes which are facultative can make use of free oxygen.

Anion

Negatively charged ions are called anions.

Annuals

These plants live only for one year. They complete their life cycle in one year, germinating from seeds and perpetuating themselves only as seeds.

Annual Cycle

It refers to the yearly cycle in plants. The plants having annual cycle complete their life in one year. They germinate from seed, flower and die perpetuating only as seed, all in the same year.

Antagonism

A relationship between organisms in which one partly or completely inhibits the growth of the other is referred to as antagonism. The two photochemical acts can exhibit antagonistic effects, one tending to oxidize cytochrome and the other reducing it.

Anther Culture

The aseptic culture of anthers on a suitable medium to produce haploid callus/plantlets is anther culture. On diploidization (using colchicine treatment or some other method) they can give rise to homozygous diploids.

Anthocyanidins

These are sugar-free counterparts of anthocyanins and identified based on the structure of a large group of polymethine dye, the benzopyrylium (chromemylium) ion. They are salt derivatives of 2-phenyl chromemylium cation also known as flovylium cation.

Anthocyanins

These are water soluble vacuolar pigments belonging to class of molecules called flavonoids. They are synthesized by organisms in plant kingdom and have been observed to occur in all tissues of higher plants including leaves, stem, roots, flowers and fruits. They also act as oxygen quenchers under stress conditions.

Antiauxin

As the name indicates they are anti to auxin transport. These are inhibitors of polar auxin transport and include substances, like 2, 3, 5 triidiobenzoic acid (TIBA).

Antibiotics

The substances used against microorganisms and are chemotherapeutic agents. They have anti-bacterial, anti-fungal or anti-parasitic activity. Antibiotics are wholly or partially derived from certain microorganisms and are used to treat bacterial or fungal infections. They act either by killing microorganisms or stop them from reproducing thus reducing their number and allowing body's natural defense system (immune system) to work for their elimination.

Antibody

Antibody forms part of the immune system. These are the proteins synthesized in response to foreign agent. Antibody or immunoglobin can bind with antigen.

Antigen

A substance which can induce an immune response and has determinant group which interacts with specific receptors or antibodies is antigen.

Antitranspirants

The substances which reduce transpiration (and usually also photosynthesis to varying degrees) by causing stomatal closure, e.g., phenyl mercuric acetate (PMA) or by forming a film of plastic or waxy materials or by increasing reflectance, e.g., application of kaolinite are antitranspirants.

Apical Dominance

The phenomenon in which growth of terminal bud inhibits or controls the growth of lateral buds is apical dominance. Decapitation of apical buds leads to the development of lateral buds. The ratio of cytokinin to auxin has been implicated in the control of apical dominance.

Apical Meristem

It occurs at the apex of roots and shoots and is responsible for the increase in length. In fact, all primary tissues of plants originate from apical meristem of roots and shoots. Shoot and root meristems divide continuously to give rise to new cells that form new organ.

Apices

The tips of root and shoot are known as apices. It is the part of plant where apical meristem is present that is responsible for growth of root and shoot.

Apoplast

It is the 'dead' part of the plant outside the plasmalemma. It consists of cell wall, intercellular spaces and water-filled xylem elements. Apoplast is utilized for transport of materials through the plant. The ascent of sap takes place in apoplast which excludes the casparian strip in endodermal cell.

Apoprotein

A polypeptide chain or protein component of conjugated protein is apoprotein. In phytochrome the apoprotein and its chromophore make up the holoprotein.

APS sulphotransferase

It is an enzyme involved in sulphate assimilation and it transfers the sulphate group of APS to the sulphur atom of an acceptor molecule.

Aquaporins

The integral membrane proteins forming water selective channels and facilitating water movement across the membranes are aquaporins.

Arrhenius Plot

It shows the relationship between temperature and physiological process of an organism. It is plotted between a physiological process occurring in an organism and the increase in absolute temperature. If an organism can live in a range of temperature for a long period the curve is linear but if it is sensitive to temperature then the curve is non-linear.

Ascent of Sap

It is the movement of water and minerals from roots to leaves. It occurs through xylem. Water from root is transported because of root pressure and transpiration pull because of cohesive and adhesive forces between water molecules and xylem.

Assimilates

They are products of assimilation and metabolism.

ATP sulphurylase

It is an enzyme associated with sulphur assimilation and reduction in plants.

Autoradiogram

To know the location of a radioactive material in a substance autoradiogram is prepared. Autoradiogram is a photographic record of the radioactive material in a substance.

Autumn Syndrome

It represents the effect of short days of autumn on plants. It includes leaf abscission, reduced stem elongation, reduced chlorophyll production, and increased formation of other pigments. It causes dormancy and leads to development of frost hardness.

Auxin

Auxins are growth promoting phytohormones. Auxins are chemical substances that increase the growth of stem or coleoptile section. They are transported from the apex and are present only in actively growing tissues like root and shoot apex. The example of a naturally existing auxin is indole-3-acetic acid (IAA). Auxin causes apical dominance by transporting itself from apex and acting as inhibitor of lateral bud growth. Auxin causes cell elongation, root formation, promotes cell division in cambium, prevents formation of abscission layer and even induces parthenocarpic fruit development.

Auxin Transport

Auxin transport is polar in nature. Auxin is transported downward from shoot tissue and upward from root. Such a polar transport may occur either because of difference in electric potential from apex to base or because of electrochemical gradient set by cellular accumulation of auxin. The transport of auxin is a slow process and is an energy requiring process. Under anaerobic conditions and because of metabolic poisons the polar transport of auxin is inhibited. In general its transportation occurs through parenchymatous cells.

Available Soil Moisture

The water retained in the soil between field capacity and permanent wilting percentage is referred to as available soil moisture.

Avoidance

One method for plants to cope with stress is avoidance. In this process organisms adopt adaptive measures that help them to reduce the severity of stress factor. They avoid direct contact with stress factors.

B

Back Cross

Genetic cross between a hybrid and one of the parents is back cross.

Bacteroids

These are large non-motile bacteria present in the infected thread of the root hair formed in the process of nodule development. They cause the inner cortical cell and pericycle cell to divide. They occur in groups in the cytoplasm of nodule. The group is surrounded by peri-bacterial membrane and between the two membranes is the peribacteriod space. It is the bacteriod in which nitrogen fixation occurs. N_2 is reduced to NH_4^+ and the electron and energy are obtained through the oxidation of carbohydrate transported to the bacteriod through phloem.

Bars

Water potential and stress level in the water relations of plant and soil are expressed in terms of pressure:
25 bars = 24.67 atmospheres

Bars of Sanio

Thickening of intercellular material and primary wall along the upper and lower margins of a pit pair are bars of sanio.

Base Response

It refers to the sensitivity of base, the response of portion below the tip. The tip is more sensitive, but the portion below that also responds because of the transportation of substance from tip to the base. Darwin observed in oat coleoptiles that the phototropic sensitivity occurs below the tip though to a lesser extent.

Batasin

They act as growth inhibitors and are obtained from yam plant (*Discorea batatas*) and cause dormancy in the bulbils. Bulbils are the vegetative reproducing structures.

Beneficial Elements

The elements which are not essential for higher plants but plants are benefitted by their presence, e.g., Na, Si, Se and Co.

Betalains

These are glycosylated compounds that are responsible for the red colour of beet root and *Bougainvillea* flowers. The red pigment in them is not the anthocyanin but they are batalains which contain nitrogen and their synthesis is promoted by light.

Betaxanthins

These are yellow coloured betalains different from anthocyanin like betalains. Betaxanthins are present in plants in which anthcyanin is absent. They are synthesized in light and are helpful in pollination.

Biennials

These plants live for two years. They remain vegetative in the first year and flower in the second year.

Bioassay

It is the biological assay in which an organism or other living system is used for studying the effect of a substance on the living organism. They take advantage of the extreme sensitivity and specificity of the organism to that substance. The bioassay may be quantified but its physical effects have to be studied. It may be quantitative when the concentration of substance is determined by the measurement of its biological response.

Biological Clock

It is a clock that works in living beings and shows periodicity in its response. It refers to the periodic oscillations that appear to be controlled by an internal time measuring system. Photoperiodism for example follows that biological clock. For biological clock the rhythm should persist under constant conditions.

Biological Magnification

The increasing concentration of non-degradable substances at each stage in a food chain is referred to as biological magnification.

Biological Nitrogen Fixation

The process involving reducing dinitrogen to ammonia by living organisms is referred to as nitrogen fixation. Bacteria like *Azotobacter* and *Azospirillum* carry out nitrogen fixation under aerobic conditions while mostly the process is carried out under anaerobic or at very low oxygen partial pressure by nitrogen fixers like *Clostridium.* Several symbiotic nitrogen fixing associations between different species of bacteria and leguminous plants are well known. Even non-leguminous plants like *Casuarina* are involved in certain symbiotic association.

Biological Strain

When a plant is subjected to stress, i.e., unfavorable environmental conditions its normal functioning is altered or rather reduced. Thus reduced or changed function is called the biological strain. The strain is said to be elastic if the change in organism is temporary and it returns to its optimal level on removing the stress. If there is permanent change in function, the strain is said to be plastic.

Biological Stress

It refers to the adverse effect on an organism, i.e., normal functioning of the organism is disturbed. It may be defined as adverse changes in plant growth and development with the change in environmental conditions. It changes the physiology of plants by affecting their normal functioning. Stress may be fatal but plants generally possess avoidance (avoid stress by adaptive measures) and tolerance (adaptation and adjustment) mechanisms.

Bioluminescence

It is the rhythmic emission of light from granular cells when they are tapped. Bioluminescence follows circadian rhythm with the peak of rhythm occurring at midnight.

Biome

Different communities (plant and animal) sharing common climate is biome.

Bioremediation

Fungi and other microbes degrade hydrocarbons like oils to CO_2 and H_2O using aerobic respiration. Such activities are referred to as bioremediation. Fungal propagules are added to contaminated land for bioremediation.

Biosphere

Sum total of all organisms and their habitats is called biosphere.

Biotic Potential

The rate at which a species can reproduce under ideal conditions is biotic potential.

Blackman Curves

According to the Blackman's principle of 'Limiting Factor' the rate of a process involving many factors depends upon the slowest factor. When many factors are involved and all are kept constant then the rate of the process depends upon the factor that is being considered. This principle was discussed in relation to photosynthesis. The rate of photosynthesis increases with the increase in the factor being considered (e.g. light) until a saturation level is reached. At this saturation point the increase in another factor becomes limiting (e.g. CO_2). Now if this limiting factor is increased, the rate of process further increases and a different saturation point is achieved until third factor (e.g. temperature) becomes limiting. If this limiting factor is increased then again the rate of process increases. A plot of curve between rate of process and a factor (e.g. light intensity) with increasing other constant factor at saturation level of first factor gives the Blackman curve.

Blotting

The method usually comprises the transference of biological molecules from a gel to a membrane filter where they can be tested for the presence of specific sequence by probes.

Bolting

It is the phenomenon of extensive internode elongation just before flowering in such a plant that does not flower in the rosette form. Low temperature and photoperiod triggers bolting. Gibberellic acid (a plant hormone) also induces bolting.

Bordered Pits

Found in dead tracheids and vessels, such pits have their primary wall swollen in the centre of pit to form torus and the secondary wall extends over the center of pit. They, thus form a boundary around pits.

Bound Auxin

Auxins complexed with cellular receptor sites and are non-diffusible are known as bound auxins. Bound auxins are bound in the cell and cannot diffuse unless extracted with solvent or hydrolyzed under alkaline condition. Infact, these are IAA with chemical conjugates like glycosyl esters or peptides. The site of synthesis of these auxins is not understood. However, these are reported in the seeds of *Zea mays*.

Boundary Layer Resistance

It is the time taken by water vapour to diffuse through unit distance across the air layer surrounding the stomata.

Brassin

Brassins are growth promoting complex mixture of lipids, isolated from *Brassica napus* and cause elongation, curvature and splitting of internodes in the bean second internode.

Brassinolide

These are plant growth-promoting steroidal lactones isolated from rape (*Brassica napus* L.) pollen. It has remarkable activity in the bean second internode assay and is the first bioactive steroid with high plant growth promoting activity. These are active in macromolecular concentration.

Brassinosteroids

The active substances isolated from pollen, stem, leaf and flower of a variety of species of *Brassica*. Brassinolide is one brassinosteroid isolated from rape (*Brassica napus*) pollen. Like auxins, it is active in micromolar concentrations stimulating elongation of hypocotyls and epicotyls of a variety of legumes as well as coleoptiles of wheat. A number of studies on mutations implicate brassinosteroids in more important role in plant development enough to consider them as a distinct class.

Brownian Movement

It refers to the movement that occurs in colloidal solutions. It is the random motion of colloidal particles caused by uneven bombardment of these particles with the molecules of the dispersion medium.

Bulk Flow

Bulk flow refers to the movement of atoms or molecules in group. Such a flow occurs in response to the difference in pressure. It is also known as mass-flow. Bulk flow is responsible for movement of photosynthates from leaves to root, based on difference in suction pressure. Munch in 1931 gave the mass flow or pressure flow hypothesis in which two osmometers were responsible for mass flow of solutes.

Bulliform Cells

These are thin-walled motor cells present in leaves. These cells do not have cuticle and so water evaporates faster in these cells than other epidermal cells. Thus, transpiration is observed to be faster in these cells than other cells.

C

C-3 Plants

They are called C-3 species because the first stable product of carbon fixation is a 3-carbon compound, 3 phosphoglyceric acid (3PGA). C-3 plants use Calvin cycle for CO_2 fixation. Most aquatic plants, trees, shrubs and temperate herbaceous plants are C-3 plants.

C-4 Pathway

It is an alternative pathway for CO_2 fixation during photosynthesis observed in C-4 plants. C-4 plants have Kranz anatomy, i.e., two different types of cells occurring in a concentric manner. Therefore, C-4 pathway involves two types of cells: the mesophyll cell and the bundle sheath cell. In the mesophyll cell initial CO_2 fixation occurs in the presence of phosphoenol pyruvate carboxylase to form oxaloacetate. Oxaloacetate is converted to malate which is transported to bundle sheath cells. Here malate is decarboxylated to pyruvate and the released CO_2 enters C-3 cycle.

C-4 Plants

These are called C-4 species because the first stable product of CO_2 fixation is a 4-carbon compound i.e., oxaloacetate. C-4 plants have Kranz anatomy i.e. they have two distinct types of cells occurring in concentric ring, the outer mesophyll cell and inner bundle sheath cells. In C-4 plants, both C-4 and C-3 cycles operate. Initially CO_2 fixation occurs in mesophyll cells and products are transported to bundle sheath chloroplast where decarboxylation takes place and CO_2 enters into Calvin cycle.

Callose

These are polysaccharides composed of glucose residues. Callose is involved in sieve plate formation. During mechanical stress callose formation occurs and used in the healing of damaged tissues.

Calmodulin

These are calcium binding proteins found in the guard cell in plants and in all eukaryotic cells. One molecule of calmodulin binds with four Ca^{2+} ions and changes its conformation. Such a change in conformation helps it in activating other proteins. Calmodulin is involved in cyclin nucleotide transport, Ca^{2+} ions transport and it also binds with enzyme groups in some phosphates.

Calvin Cycle

It is also called photosynthetic carbon reduction cycle. It occurs in the stroma of chloroplast and consists of 3 main parts; carboxylation, reduction and regeneration. During carboxylation CO_2 is fixed with ribulose, 1-5 bisphosphate in the presence of ribulose, 1-5 bisphosphate carboxylase/oxygenase (Rubisco) to form two molecules of 3 phosphoglyceric acid (3-PGA). In reduction phase, 3-PGA is reduced to glyceraldehyde-3-phosphate and during regeneration ribulose, 1-5 bisphosphate is formed.

CAM Plants

They are mostly succulent and can grow in dry habitat. CAM plants have a unique feature of stomatal functioning. Their stomata remain open during night and close at day time. This feature enables CAM plants to conserve water. As the stomata are closed during day the water lost through transpiration is minimized. The open stomata during night fix CO_2 and malate is formed that is decarboxylated to evolve CO_2 during day time.

Cambium

Cambium is present between xylem and phloem in vascular bundles and are called vascular cambium. Cells of cambium are thin walled, vacuolated and divide periclinally during secondary growth to form secondary xylem and phloem. Cambium is composed of two types of cells:

i) Fusiform initials that are long with tapering end giving rise to secondary xylem and phloem and (ii) Ray initials which are isodiametric cells giving rise to secondary medullary rays. Cambium is absent in monocots and so they do not show secondary growth.

Capillarity

It is the adhesion between a solid surface and a liquid and it helps in upward rise of water in a capillary tube. Capillarity may not be responsible for ascent of water in plants but to a smaller extent it plays a role in holding water in some parts of plants, like in the cell wall of leaves and other tissues.

Capillary Water

After rainfall some water percolates down under the influence of gravity but a large amount of water is retained in the interstitial spaces in the form of thin film around the soil particles. This is known as capillary water and it is easily available to plants. Infact plants absorb capillary water.

Carbohydrates

These are the major building blocks of cells made up of small sugar units. They have the general empirical formula $(CH_2O)n$ where n is 3 or greater. The formula shows that they are hydrates of carbon. Carbohydrates are the major organic molecules synthesized during photosynthesis from which all organic foodstuffs are derived. They are of three types.

i) Monosaccharide is the simplest carbohydrate with only one aldehyde or ketone as a functional group.
ii) Oligosaccharides have 2-10 monosaccharide units joined by glycosidic bond.
iii) Polysaccharides have multiple monosaccharide units and are the major carbohydrates involved in both structural and functional aspect of plant cell.

Carboxyl Group

These are –COOH containing group that impart acidic character. They are constituents of amino acid.

Carboxylation

It refers to the addition of CO_2 to a substrate. This process is observed in the Calvin cycle that occurs in the stroma of chloroplast. CO_2 is added to RuBP to form two molecules of 3-PGA. Thus, carboxylation results in the fixation of carbon.

Cardiac Glycosides

These are sugar containing sterols that have allelochemical activity and cause heart attack in vertebrates. Since they are concerned with heart attack they are called cardiac glycosides. Though harmful in this respect they are used medicinally to strengthen and slow heart beat during heart failure. An example of cardiac glycoside is degilamides found in germs digitalis. Although it is toxic to heart as it blocks the Na-K ATPase of heart muscle membrane, it is still used as a cure against heart failure.

Cardinal Point

In the process of photosynthesis each factor has minimum, optimum and maximum value. These three values are the cardinal points. When a graph is plotted between photosynthesis and limiting factor the points at which minimum, optimum and maximum values are obtained represent the cardinal points.

Cardinal Temperature

Growth of a plant is affected by minor variation in temperature. Each species or variety responds differentially to temperature changes. They have minimum temperature below which they cannot grow, optimum temperature at which growth is maximum and maximum temperature above which they cannot grow. If other factors affecting growth are not limiting then cardinal temperature generally defines the temperature range over which growth is possible. However, this temperature varies depending upon region where species grow because each species is adapted to its own natural temperature environment.

Carotenes

These are orange or red-orange coloured pigments present in carotenoids family together with xanthophylls. They are pure hydrocarbons containing carotenoids. Among various carotenes, β-carotenes protect chlorophyll from photo oxidative damage.

Carotenoids

These are group of isoprenoids that perform important functions in plants. They act as accessory pigments absorbing wavelength between 400-500 nm and transfer it to chlorophylls. Carotenoids protect the photosynthetic apparatus from photo-oxidation by acting as quenchers of singlet oxygen species. Carotenoids are of two types: the pure hydrocarbon-containing carotenes and the oxygen-containing xanthophylls. They are considered to be precursors of abscisic acid.

Carrier Proteins

These are a class of transport proteins that transport solute and ions across the membrane. When a solute comes in contact with the carrier protein, the protein undergoes conformational changes and transports the solute to the other side of the membrane, and thereafter, reverts to its original shape, ready to pick up another solute. Carrier protein's function is obvious during the process of facilitated diffusion.

Carry Over Effects

It refers to the effects of environmental factors on growth that are carried from one generation to another under adverse conditions. Plant growth is retarded in successive generation. If seed from such plants is` taken and grown under optimum environment needed for growth it will still take two or more generation to come to its normal growth. This effect on growth condition is thus carried from one generation to another and is known as carry over effect. This effect does not violate genetics because it is not the genes that carry this effect but the embryo and stored food in cotyledon or parent plant itself are conditioned in such a way to carry over the effect of adverse environment.

Carrying Capacity

Maximum number of organisms that an environment can support is carrying capacity of the environment.

Casparian Strips

These are present in the radial and transverse walls of root endodermal cells. They are made of suberin, a complex mixture of hydrophobic molecules, long-chain mixture of fatty acids and alcohols. The hydrophobic molecules impregnate the cell wall filling the spaces between the cellulose microfibril as well as intercellular spaces between cells and thus act as barrier in the apoplastic movement of water in the endodermis.

Catabolism

A process that results into the breakdown of large molecules into smaller ones, releasing energy in cells, e.g., oxidative breakdown of sugars or the process of respiration in which glucose is broken down into CO_2 and H_2O.

Catalytic Protein

The enzyme involved in catalyzing the hydrolysis of ATP.

Catechols

Soil siderophores are produced by microbes and are iron-chelating agents. They are phenolic compounds and exist chemically as catechols or hydroxamates.

Cation

Positively charged ion is called cation and is formed by the loss of an electron from the metal.

Cell Cycle

It refers to a complete cycle of a new cell formation from a pre-existing cell. A newly formed cell undergoes growth and division to form new daughter cell. The cell cycle is divided into two stages: interphase and mitosis. Interphase is the longest phase where synthesis of DNA occurs. In fact, interphase prepares the cell for division. It is divided into G1, S, G2 and it is followed by mitotic phase. Cell cycle includes all the four stages. In G1 phase, synthesis of RNA and protein occurs. It is followed by S phase where synthesis of DNA occurs. In G2 phase protein required for spindle formation is synthesized. It is followed by M phase where the prepared cell is ready to divide into two daughter cells. It includes prophase, metaphase, anaphase, telophase and cytokinesis.

Cell Division

It is the process of formation of daughter cell from parent cell. A cell divides when it has attained certain maximum size which disturbs its karyoplasmic ratio. Cell division is essential for cell growth and multiplication. On the basis of the manner in which nucleus divides, cell division is of four types, i.e., (i) Amitosis: in which nucleus elongates and constricts followed by constriction of cytoplasm to form two daughter cells, e.g., *Chara* (algae), yeast, endosperm or diseased plant part (ii) Mitosis: in which there is equational division and two daughter cells formed have same number and type of chromosome. It occurs in somatic cells (iii) Meiosis: it is reductional division occurring in reproductive cells responsible for reducing the number of chromosome in daughter cell to half of the parent cell and (iv) Free nuclear division.

Cell Membrane

The cell is surrounded by a membrane that lies at the periphery of cytoplasm and is called cell membrane or plasma-membrane. It is a differentially permeable membrane responsible for transport of solutes in and out of cell. It also helps in translation of external stimulus to biochemical and physiological responses and is involved in the synthesis and assembly of cell wall microfibrils. Membrane is made up of lipids, proteins and carbohydrates. Lipids include phospholipids, glycolipids, sterols and neutral lipids. Proteins are either integral (intrinsic) or peripheral (extrinsic). Carbohydrates include glycolipids, glycoproteins and various cell wall polysaccharides. The cell walls also contain enzymes like K^+-ATPase, cellulase, glyceryltransferase.

Cell Plate

It is present in between two dividing daughter cells in plants. It is the initial partition between cells during cytokinesis. Cell plates are formed by fusion of golgi vesicles which deliver the middle lamella precursors.

Cell Wall

It is a semi-rigid, non-living wall present outside the plasma membrane in plant cell. It is mainly composed of cellulose, hemicelluloses, protein, pectin and lignin. It is mainly of two types: primary cell wall and secondary cell wall.

Primary cell wall is formed around differentiating plant cell while the mature cells have secondary wall. The main function of cell wall is to provide strength and rigidity to the plant cell. Cell wall helps in growth of the plant cell.

Cellular Differentiation

It refers to the specialization that occurs in a cell. The cellular differentiation follows cell division and cell enlargement. Through differentiation meristematic cells give rise to specialized structure.

Cellulose

It is a structural polysaccharide in plants formed from D-glucopyranose units linked by β-(1-4) glucosidic linkage. It is the major component of cell wall. It is insoluble in water, chemically inert and lacks sweetness. Cellulose is digested by enzyme cellulase that is absent in man but present in many animals.

Chelates

These are stable complexes that are formed between metal ion and an organic molecule (ligand or chelating agent). A coordinate bond is formed between metal and ligand by sharing of electron pairs between them. In such a complex the metal ion can be easily removed and the plant can easily take them. Thus the formation of chelates is helpful in translocation of nutrients that form insoluble precipitate, e.g., iron.

Chemical Potential

It is the free energy per gram molecular weight (i.e. per mole) of any substance.

Chemiosmotic Theory

A phenomenon that couples electron transport chain with ATP synthesis is known as chemiosmotic theory. It was given by Peter Mitchell. Electron transport through respiratory chain in the mitochondria results in the development of an electrochemical H^+ gradient across the inner mitochondrial membrane. The electrochemical potential of this gradient is utilized in the synthesis of ATP. For net transport of protons across the membrane the respiratory chain and ATP synthase are placed vectorially.

Chilling Injury

It refers to the injury caused to plants due to low temperature. Chilling creates a stress on plants affecting its metabolism. It affects plants growing in tropical areas. Freezing temperature acts on the membrane lipid content and changes lipid from liquid to crystalline state and thus affects its function. This change depends upon the ratio of saturated to unsaturated fatty acids. On increasing the proportion of unsaturated fatty acids the membrane remains functional. Chilling stress increases the permeability of the membrane and therefore affects enzyme activity and other metabolic functions.

Chlorophylls

These are green coloured photosynthetic pigments involved in photosynthesis and thus present in all autotrophs. It has a tetrapyrrole (porphyrin) ring containing magnesium joined to a long-chain alcohol phytol. Chl *a* and Chl *b* are the two main types of chlorophyll pigments present in higher plants. In the process of photosynthesis chlorophyll collects light quanta or it generates free electrons leading to the production of reducing compound and ATP. It is also involved in processes like photoaxis and photokinesis both of which require light for motion.

Chlorophyll *a* (Chl *a*)

It is present in all green plants and has the formula $C_{55}H_{72}O_5N_4Mg$. In addition to the pyrrole ring it has a cyclopentanone ring with a methyl ester. A phytol ester is linked by a propionic acid with the pyrrole ring D and it has a formyl group at C7 instead of a methyl group. It has a methyl group at C3.

Chlorophyll *b* (Chl *b*)

It is yellow-green in colour and has the empirical formula $C_{55}H_{70}O_6N_4Mg$. Its structure is similar to Chl *a* except for the replacement of the methyl group at C3 by an aldehyde group.

Chloroplast

It is a cytoplasmic, semi-autonomous organelle where photosynthesis occurs. It has three major regions:

i) The outer envelope consisting of double membrane.
ii) Stroma which is a hydrophilic, proteinaceous matrix consisting of enzymes, and ribosomes and in higher plants and algae even DNA is present. In stroma the dark reaction of photosynthesis occurs.
iii) Highly organized internal lamellar membranes containing photosynthetic pigments. These membranes stack together to form granum. In the grana light reactions of photosynthesis occur. Thus the membranes are also used in energy capture. In the granum each disc is called thylakoid. Thus granum is a stack of thylakoids. In the granum thylakoids are connected by intergranal or stroma lamellae.

Chloroplast DNA

DNA of the chloroplast is double stranded, circular and organized like that of the prokaryotes and coding for some but not all chloroplast proteins.

Chlorosis

It refers to the physiological state that occurs due to low chlorophyll synthesis and results in pale green or yellow coloration of leaves.

Chromatin

It is present in the nucleoplasm of nucleus and is a network of threads. Chromatin is made up of DNA and protein. It is a network of threads in the initial phase of mitosis but as mitosis proceeds it becomes condensed and forms chromosome.

Chromophore

It is a prosthetic group attached to the phytochrome protein and helps the phytochrome to respond to light by absorbing light itself. It is an open chain tetrapyrrole. On absorption of light the structure of chromophore is altered and this in turn affects the phytochrome structure which is responsible for its physiological activity.

Chromoplast

These are plastids containing carotenoid pigments. Chromoplast gives colour to leaves, flowers, and fruits. In ripening fruits chloroplast changes to chromoplast as in tomato.

Circadian Rhythm

Such rhythm occurs in living organisms and is related to the daily changes that occur in the physiological and metabolic activities in higher plants. Since these rhythms are not affected by the daily cycle of light and darkness and other external factors, these are called endogenous circadian rhythms. Flowers and plant processes like photosynthesis, respiration, leaf movements, stomatal opening, etc. follow circadian rhythms.

Clone

Genetically identical cells or organisms produced from a common ancestor by asexual means.

Cochromatography

Co means together. It is the chromatography of two samples together. This method helps in identification of an unknown substance by comparing it with the chromatograph of a known substance.

CO_2 Compensation Point

It is the atmospheric concentration of CO_2 at which the rate of photosynthesis is just equal to rate of respiration. This point is reached when the amount of CO_2 taken up in photosynthesis is equal to that generated during respiration. Apparent photosynthesis at this point is zero.

Coenzyme

A complete enzyme has two parts, coenzyme and apoenzyme. Coenzymes are small, organic non-proteinaceous parts of an enzyme needed for the catalytic activity of the enzyme. These molecules (coenzymes) are consumed during the course of reaction but are recycled continuously by another enzyme. Coenzymes perform various functions like the transfer of groups (e.g. ATP) or oxidation-reduction reactions.

Cohesion

It is the force of attraction between like molecules. Cohesion is responsible for ascent of sap. During transpiration water evaporates from the leaves and water from roots moves to leaves due to cohesive force between water molecules in the xylem vessel.

Cohesive Force of Water

Two water molecules are held together by H-bonds. This H-bond or the force between two water molecules is called the cohesive force of water. This force helps in forming droplet of water instead of spreading out. It is the cohesive force of water that is responsible for the ascent of sap (cohesive theory).

Colchicines

These are the chemicals which induce polyploidy in plant cells by inhibiting chromosome segregation during meiosis.

They act as spindle poison during cell division. They bind the tubulin protein of microtubules and thus inhibit its polymerization.

Colloids

These are small particles that do not settle out by gravity. They are larger than true solute particles. Colloids remain suspended in water because the water molecules are small and in constant motion. As a result of this collision colloidal particles are never at rest and therefore do not settle down. Colloids exhibit many important properties like tyndall effect, Brownian movement etc.

Companion Cells

These are parenchymatous cells accompanying the sieve tube members. They are derived from the same mother cell as the sieve-tube member and have various cytoplasmic connections with it. Companion cells provide metabolic support to sieve-tube members and are functional as long as the sieve-tube members are functional. Such an intimate connection and location of these cells give them the name companion.

Competitive Inhibition

They are structurally very similar to the substrate of an enzyme and can combine with the enzyme at its active site thereby decreasing the activity of the enzymes. This type of inhibition can be removed by increasing the concentration of substrate that can displace the inhibitor and can itself bind at the enzyme's active site.

Conduction

It is the transfer of energy between two surfaces, from a hot one to cold one till both reach the same level of energy.

Conjugated Auxin

Such auxins have their carboxyl group covalently bound to other molecules. The bound auxin can be removed by hydrolase enzyme.

Constitutive Enzyme

This is an enzyme that is present in approximately constant amounts in a system irrespective of the fact, whether its substrate is present or absent.

Convection

It is the cooling of air being warmed. Warm air becomes light, expands, rises up, cools and then comes back to its site of origin.

Criteria of Essentiality

This was given by Arnon and Stout in 1939. An element is said to be essential if it fulfils the conditions that (i) plant cannot complete life cycle in its absence, (ii) it cannot be replaced by other elements, and (iii) it has a direct effect and its effect is not the result of synergism or antagonism.

Crop Growth Rate

Dry weight accumulation (or energy content) per unit area in unit time interval is crop growth rate.

Cryptochrome

These are blue light photoreceptors also called 'hidden pigment' as their exact identity is not clear. However, carotenoids and flavin may be called cryptochrome for blue-light response as their absorption spectra match the action spectra for phototropism. This similarity in the spectrum is considered because phototropism shows blue light response. Beside phototropism, inhibition of hypocotyl elongation, stomatal movement, pigment biosynthesis and gene activation also show blue-light responses in the wavelength range of 400-500 nm. The action spectrum of blue light response is a "three-finger" structure in 400-500 nm wavelength.

Cumulative Water Loss

It includes the amount of water lost through transpiration and evaporation both.

Cutin

Cutin is a complex mixture of fatty acids with some hydroxyl group attached along with hydrocarbon chain.

Cyanide Resistant Respiration

The respiration that continues even in the presence of inhibitors of cytochrome oxidase like cyanide and azide which combine with iron in cytochrome oxidase preventing electron transport and poisoning respiration. This type of respiration is found in the spadix of aroids like *Arum maculatum*, inflorescence of *Sauromatum gutlatum*, flowers of *Symplocarpus foetides*, *Philodendron selloum* and in several higher plants and microorganisms. Cyanide-resistant respiration is possible because of an alternative path in the mitochondria for electron transport besides the main pathway in which cytochrome oxidase is blocked. The alternative route branches from main respiratory chain in the ubiquinone region and allows electron transport to oxygen from ubiquinone to a flavoprotien to an alternative oxidase. In such respiration heat is produced instead of ATP. The alternative pathway is however sensitive to SHAM (salicyl hydroxamic acid) and CLAM (chlorbenzhydraxamic acid) that inhibit alternative oxidase.

Cyclic Electron Transport

As the name indicates it involves the cycling of electron from reaction center P_{700} back to P_{700} in a definite way passing through various carriers. Here the P_{700} is excited and releases $2e^-$ that is accepted by Fd→Fd-PQ oxidoreductase→Cyto b_6→PC and recycled back to P_{700}. In the same process simultaneous to e^- release two H^+ are deposited in the thylakoid channel when PQH_2 is oxidized. In this transport water reduction does not occur as PSII is not involved, NADP is not reduced but only ATP is synthesized.

Cyclic Photophosphorylation

It refers to the formation of ATP by the cyclic electron transport pathway. ATP is produced by coupling factor in response to the decreased pH in the thylakoid channel. This decrease in pH is because of more H^+ ion inside the thylakoid channel and less H^+ outside. Such a difference in H^+ results in a proton motive force that supplies energy for the phosphorylation of ADP with Pi forming ATP.

Cytokinesis

It is the division of cytoplasm that follows mitosis and divides the cell into two equal daughter cells.

Cytokinin

It is a phytohormone produced in meristematic region and helps in cell division. Besides this it also helps in cell enlargement, root growth and lateral root initiation, bud development, shoot growth and delays senescence.

Cytoplasm

It is the liquid portion of the protoplasm outside the nucleus present in all cells. It is bounded by plasmalemma. Cytoplasm contains cytoplasmic organelles like mitochondria, plastids, ribosomes, microtubules and microbodies. It also has membranous structures like endoplasmic reticulum and golgi bodies.

Cytoplasmic Streaming

It is the movement of cytoplasm that is responsible for solute translocation. Solute is translocated through phloem that contains sieve tube elements. In the process of cytoplasmic streaming solute particles move from one cell of the sieve tube element to another under the influence of circulating cytoplasm. The solute diffuses through cytoplasmic strands connecting one element to another and hence translocated.

Cytosol

It is a colloidal solution of organic substances present within the protoplasm but outside all other cell organelle and membrane system. Cytosol contains many dissolved enzymes and many important functions are performed in this compartment.

D

Day Neutral Plant

Plants of this group are not affected by day length and flower the whole year. Examples are tomato, cucumber, cotton, etc.

Decarboxylation

It refers to the reaction in which CO_2 is removed from a molecule.

Dedifferentiation

It is the process in which cell reverts from its differentiated form to its embryonic form (i.e. undiffer-entiated or meristematic). Differentiation is observed in totipotent cells. A single specialized cell undergoes cell division to form an undifferentiated callus that eventually gives rise to a new plant.

Deficiency Disease

Deficiency disease is caused by the deficiency of essential mineral nutrients such as: Manganese: grey speck of oats, marsh spot of peas, speckled yellows of sugar beet; Boron: heart rot of beets, stem crack of celery, water core of turnip, drought spot of apples; Zinc: little leaf, rosette of apples, white bud of maize; Copper: die back; Molybdenum: whiptail of cauliflower.

Deficient Zone

Low concentration in a graph plotted between nutrient concentrations and growth represents deficient zone. An increase in the nutrient concentration increases growth.

Delayed Effect

It refers to the delay in response in some developmental process which occurs in response to some environmental factors that work as a stimulus. The plant receives the stimulus, undergoes certain changes and responds to it after some time.

Denaturation

It refers to the loss in activity of a molecule like protein, DNA etc. Denaturation occurs when the structural features of a molecule are disturbed by harsh treatments like high temperature, UV radiation, changes in pH etc. When a protein is denatured it loses its properties like solubility, specific activity and even its function. Denaturation in general cannot be reversed.

Denitrification

It is the process of conversion of nitrate and nitrite into ammonia, nitrogen gas and nitrous oxide. The end product of denitrification is nitrogen that is released into the atmosphere. Anaerobic bacteria like *Pseudomonas denitrificans*, *Bacillus subtilis*, *Azotobacter*, *Clostridium*, *Thiobacillus denitrificans* are involved in denitrification using nitrate as an electron acceptor during respiration instead of oxygen.

Desert Ephemerals

They are the xerophytes that can tolerate the drought conditions. In dry season during drought they remain as dormant seed and do not germinate. These are short-lived annuals and mature in one season. On soil moisture becoming sufficient to support their growth they germinate and mature leaving seed before the dry season approaches again.

Desmotubule

These are tubular structures present in the central region of plasmodesmata and are compressed parts of the endoplasmic reticulum extending from one cell to the other. Solutes cannot pass through it but can move between it and the plasma membrane.

Determinate

Any structure is determinate if it grows to a certain size and has a limited life span. Leaf, flower, fruit are all determinate structures because they grow to their full size and finally senesce and die.

Development

It is the process in which simple structures undergo differentiation and maturation to form complex structures. It is a qualitative process and may occur following growth but it may also occur without growth.

Devernalization

It is the reverse of vernalization. When a vernalized flower is given high temperature treatment then devernalization occurs and the effect of low temperature is removed. In general, a temperature of 35°C or above is used for devernalization. However, devernalization becomes difficult if vernalization occurs for a longer duration.

Dielectric Constant

It refers to the potential to neutralize the attraction between electrical charges. Higher the dielectric constant of a solvent better is its capacity to dissolve substances. Water has the highest dielectric constant and is therefore the best known solvent. Water forms strong bond with other molecules through attraction between unlike charges and therefore has high dielectric constant.

Differentiation

This is the process in which differences (excluding size) arise amongst cells, tissues and organs. It involves normally a number of qualitative changes. For example, an adult organism develops from a zygote as a result of cell division and cell differentiation.

Diffusion

It involves the movement of substance from the region of their higher concentration to the region of their lower concentration. Thus diffusion occurs along the concentration gradient due to a pressure developed called diffusion pressure. Pure solvent has maximum diffusion pressure which decreases on addition of solute resulting in a diffusion pressure deficit. Diffusion is a very important phenomenon involved in all physiological processes in plants. The movement of water in plant, the movement of substances in and out of cell through plasma membrane, the intake of gases and their movement through intercellular spaces of the tissues, absorption of water and minerals and processes like transpiration all occur through the process of diffusion.

Dioecious

A flower is dioecious if it contains male and female flowers on different plants. The flower may be staminate (male) or pistillate (female).

Disulphide Bond

It is formed between two cysteine molecules through interaction of sulphur present in the molecules. Disulphide bonds are present in the primary structure of protein and join two polypeptide chains together.

Donnan Equilibrium

The cooperation of both electrical and diffusible phenomenon is considered in Donnan equilibrium. It accounts for the effect of fixed or indiffusible ions.

Dormancy

It is the condition of seeds when it is non-functional (reduced activity) and does not germinate even under favourable environmental conditions. It occurs when the seed coat is very hard and is impermeable to water and gases. The hard seed coat in the absence of water does not allow germination and the hard coat also resists embryo expansion. Other reason of dormancy can be immature embryo and presence of inhibitors. Such dormant seeds develop when the embryo matures. In dormancy the seeds generally remain viable. It is a good way of seed survival for a longer time and which germinate only under most favourable conditions. However, to break dormancy either the seed coat can be mechanically ruptured or a pressure can be applied on it to increase its permeability or it may be exposed to light or given low temperature treatments.

Drought

Deficiency of rainfall resulting into depletion of soil water and consequent injury to plants is referred to as drought.

Drought Avoidance

The ability of plants to withstand periods of rainfall deficit and resulting soil water deficit by maintaining higher water status is drought avoidance.

Drought Escape

The ability of a plant to complete its life cycle before serious soil and plant water deficit develops. Traits like developmental plasticity help achieving the plants this phenomenon.

Drought Stress

It refers to the stress that is created in plants in response to scarcity of water. Drought stress affects the metabolism of plants. The membrane becomes exceptionally polar, protoplast shrinks, protein of cytosol and organelles gets denatured. Root growth is affected but the major effect is on shoot and growth of leaf.

Drought Susceptibility Index

It is a reflection of the genotype under drought stress conditions as compared to its maximum potential under irrigated conditions.

Drought Tolerance

The ability to endure water deficit under decreasing tissue water potential is drought tolerance.

Dry Matter

It is the quantitative measurement of growth in which the amount of protoplasm or dry matter is determined. In this water is excluded from measurement because water content varies depending upon water status of the plant. It is compared to fresh weight.

E

Ecads

These are plants differing in appearance but belonging to essentially homologous genetic stocks. They are also known as ecophenes. If the plants differing in appearance are grown in same environmental conditions, they will appear as same.

Ecological Niche

The functional aspect of a species is called ecological niche.

Ecotone

The zone of transition between the two communities is ecotone.

Ecotype

Same species that differ genetically but are interbreeding are known as ecotypes. These are the strains within the species adapted to specific environment.

Ectomycorrhizae

Mycorrhiza is the association of fungi with plant root. Such an association represents mutualism in which both partners are benefitted. Ectomycorrhizae is a condition described as presence of fungi externally on the root and not inside the root. They are short, highly branched and covered by tightly interwoven mantle of fungal hyphae. Although the fungus is outside, it penetrates the intercellular space of root cortical cells forming the intercellular network called Hartig net.

Elastic Biological Strain

It is the biological strain that causes a temporary change in organism's function and on the return of favourable environmental conditions the organisms return to their normal functioning.

Electrophoresis

It is the process in which charged molecules are separated in the presence of an externally applied electric field. The movement of molecules in an electric field depends upon their charge and molecular weight.

Electropotential

It refers to the difference in potential that is created across the membrane. During hydrolysis of ATP into ADP and inorganic phosphate, ATP binds with the ATPase and is hydrolyzed taking OH^- from outside and H^+ from inside. This makes the cytosol slightly negative and its pH rises. This electopotential difference attracts cations inside the negatively charged cell and repels anion. Thus it helps in absorption of solute particles irrespective of concentration but depending on energy in terms of ATP.

Embryo

It represents a miniature plant present in ovule. It is formed from zygote in the embryo sac of ovule. The embryo contains a suspensor that helps to absorb nutrients from the surrounding endosperm, radicle that gives rise to root, and plumule that forms the shoot. It also contains cotyledon, two in case of dicot plant and one in monocot plant.

Embryogenesis

It is the process of formation of embryo from callus (unorganized mass of cells). The embryo thus formed, forms root and shoot and gives rise to a young plant. For embryogenesis to occur both cytokinins and auxins are needed. Cytokinins helps in shoot formation and auxins in root initiation.

Emerson's Enhancement Effect

Emerson and his co-worker studied the effect of two superimposed beams of light on the rate of photosynthesis. It was found that when wavelength greater than 680 nm was used photosynthesis decreased (this decrease was measured in terms of quantum yield). But this decrease was overcome by the simultaneous application of shorter wavelength. The effect of simultaneous application of the two beams increased photosynthesis greater than individual beams applied separately. This photosynthetic effect is called Emerson's Enhancement Effect and it shows the existence of two pigment systems in light reaction of photosynthesis.

Endodermis

It is a layer of cell present outside the pericycle. The radial and transverse wall of endodermal cell contains casparian strips that are impregnated with suberin. The endodermis offers resistance to water movement because of the presence of casparian strips that are impermeable to water. Water moves into xylem through the protoplast.

Endomycorrhizae

It is the mycorrhizal association in which the fungal hyphae grow inside the root cortical cells. The most common type of endomycorrhizae is the VAM (Vesicular Arbuscular Mycorrhizae), whose hyphae grow both within and between the root cortical cells forming a branched structure. VAM helps to increase the surface area for nutrient exchange between host plant and the fungus.

Endoreduplication

It is the process in which chromosomes duplication takes place without cell division. In mitosis chromosomal DNA replicates but when this division is not followed by cytokinesis then instead of dividing equally into two daughter cells it remains in a single cell increasing its ploidy level. Endoreduplication is widespread in plants and helps probably in differentiation of post-mitotic cells. A good example of this process is observed in polytene chromosome where chromosome duplicates without cell division.

Endosperm

It is the primary nutritive tissue. An embryo derives its nutrition from endosperm through its haustoria. Endosperm is generally triploid formed from primary endosperm cell (a product of triple fusion, i.e., male gamete + secondary diploid nucleus). Endosperm is of three types.

i) Nuclear: In this type free nuclear divisions occur and the nuclei arrange at the periphery. Cytokinesis begins at the periphery making it cellular at maturity.
ii) Cellular: In this type cytokinesis immediately follows nuclear division making it cellular from the beginning.
iii) Helobial: In this type first mitosis is followed by cytokinesis giving rise to unequal cells which undergo free nuclear divisions.

In general endosperm is triploid but in some cases they may be haploid as in gymnosperm.

Energy of Activation

It is the amount of energy required to increase the rate of a reaction. In the absence of a catalyst substrates form intermediates before products are formed. These intermediates are at higher energy level compared to substrate and therefore an initial energy of activation is needed for substrates to be converted into product. Externally energy is given in the form of heat. In contrast, in an enzyme-catalyzed reaction the enzyme combines with the substrate and lowers the energy of activation so that large number of substrate molecules can be converted to product.

Enthalpy

It is the sum of internal energy (E) plus absolute pressure (P) multiplied by volume (V) and is represented by:

$$H=E+PV$$

Enthalpy is a state of function and is taken at constant pressure. The change in enthalpy at constant pressure is equal to the heat change at this pressure.

Entropy

It is a state of disorder or randomness. It is greatest at the time of death. In a fully organized system entropy is lowest.

Environment

The sum total of all living and non-living things that are present on the earth corresponds to environment. Air, water, animal, plants and energy all are included in environment. Thus all biotic and abiotic factors that affect an organism's growth and development are its environment. An organism responds to change in environmental conditions by adapting itself morphologically or physiologically.

Enzyme

Enzyme is a biological catalyst that works to convert substrate into product without itself undergoing any change in the process. All enzymes have an active site to which a substrate molecule binds forming an enzymes-substrate complex. It is this complex that dissociates into product(s) and the enzyme is released. Enzymes have high specificity and they work by lowering the activation energy of the reaction.

Enzyme Denaturation

It refers to the loss in activity of an enzyme which occurs either at high temperature or in presence of an oxidizing or reducing agent. At such treatment the protein in enzyme is denatured and therefore enzyme loses its activity. At high temperature H-bonds are broken and covalent bonds formation may occur between different polypeptide chains and part of the same chain. An oxidizing or reducing agent acts on disulphide bridges.

Enzyme Induction

It refers to an increase in the activity of enzyme by the presence of an inducer. This induction increases the rate of enzyme synthesis and is brought about when inducer (substrate) reacts with repressor molecule and therefore prevents blocking of the operator by repressor. Thus it causes derepression of the gene that is responsible for enzyme synthesis and hence increases enzyme synthesis.

Enzyme Substrate Complex

In an enzyme-catalyzed reaction, an enzyme reacts with substrate and binds to its active site forming enzyme-substrate complex. This complex dissociates in the course of a reaction to release the product and the enzyme becomes free to accept another substrate.

Ephemeral

Plants having very short growing period (days to a few weeks) or short flowering period (usually one day or less) are referred as ephemerals.

Epicotyl

Epicotyl is the embryonic shoot within the embryo. It is present just above the cotyledon and gives rise to stem and leaf.

Epinasty

It refers to the downward bending of petiole and leaves with tips inclined to ground. It may occur due to excessive cell elongation on the adaxial (upper side of petiole). Unequal flow of auxin on the two sides of petiole or ethylene treatment may also cause epinasty.

Essential Elements

These are the elements required by higher plants for normal growth and development fulfilling the criteria of essentiality. There are 17 elements considered as essential elements. These may be macro- or micronutrients.

Ethephon

It is a compound that is used for ethylene production. It is chemically 2-chloroethyl phosphonic acid and is commercially used for ethylene production. It is easy to regulate ethylene concentration in experiments with the use of ethephon.

Ethylene

It is the only gaseous plant hormone and an important plant growth regulator. It causes fruit ripening, releases dormancy, causes the growth and differentiation of shoot and root through triple response, abscission, senescence and root formation. Ethylene is important as it causes flower induction in some plants and it also acts as a wound hormone reducing stress. Ethylene has now been recognized as an important growth regulator controlling every aspect of plant growth and development starting from germination to death.

Etioplast

When the plants are placed in dark then the chloroplast is not exposed to light. Such chloroplasts are called etioplasts. Etioplasts have precursor pigment for chlorophyll, the protochlorophyllide. In the presence of light the enzymes for chlorophyll biosynthesis become active for chlorophyll synthesis converting etioplast to chloroplast. High concentration of etioplast makes the leaves yellow.

Eutrophication

It is the enrichment of water bodies like lakes, estuaries by excessive nutrient deposition particularly nitrate and phosphate. These nutrients stimulate excessive plant growth especially of algae causing the algal bloom. The algal bloom consumes a lot of oxygen and causes depletion in dissolved oxygen (DO). Reduced DO is fatal for fish living in such water bodies causing their death. Thus eutrophication is a serious ecological problem that is caused by discharge of nutrient rich waste like sewage effluent, fertilizers runoff etc. into water bodies.

Euxerophytes

These are true xerophytes and can tolerate extreme drought conditions.

Evaporation

It is a physical process in which water moves (evaporates) into air from soil or water bodies.

Evapotranspiration

It refers to the process of evaporation and transpiration taken together. It is the evaporation of water from ground-water table and transpiration of water through leaves from groundwater table. In fact, transpiration is evaporation of water from plant leaves.

Exponential Growth

The process of increase by fixed proportion in each unit of time is exponential growth.

F

Facilitated Diffusion

This is a type of diffusion in which substance moves across the membrane from a region of higher concentration to lower concentration without the involvement of energy. The movement of substance takes place through protein present in the membrane that acts as a career and facilitates the transport of substance.

Facultative Halophytes

Halophytes are salt-tolerent plants. They grow in salty environment where the osmotic potential of soil is high and water uptake is difficult. Halophytes may be obligate that grow only in salty environment or facultative halophytes that can grow both in salty water or land (salty environment) and also in non-salty environment.

F-Duction

The transfer of genes from one bacterium to another using F-factor as a vehicle is F-duction. F-factor is fertility factor and is a plasmid comprising maleness on the cell that contains it.

Fasciations

It is a disease caused by *Corynebacterium fascious* and is characterized by formation of multiple lateral branches in various dicots. The bacteria also cause certain kind of witches-brooms in which lateral bud formation occurs giving rise to branches. In fasciations the bacteria release cytokinin which promotes bud formation and lateral branching. The cytokinin is identical to those present in tRNA of bacteria. This similarity shows that the breakdown of tRNA is responsible for fasciation.

Feedback Inhibition

It refers to the inhibition of enzyme activity due to accumulation of end product. The accumulated end product acts through allosteric transformation of enzyme, preventing it from combining with substrate.

Fermentation

It occurs under anaerobic conditions when pyruvate and NADPH are accumulated after glycolysis. The end-product of fermentation is either lactic acid or ethanol. Pyruvate is decarboxylated to acetaldehyde which is reduced by NADH to form ethanol by alcohol dehydrogenase, while lactic acid dehydrogenase converts pyruvate to lactic acid using NADH as the reducing agent. Fermentation is very useful in industries where yeast converts glucose to ethanol and this is used in the manufacture of alcohol. This process is also observed in human muscle cells where after strenuous exercise oxygen becomes limiting and lactic acid formation occurs anaerobically.

Ferridoxin Thioredoxin System

This system operates to make inactive photosynthetic enzymes active. There are many enzymes in photosynthesis that are inactive in dark condition and become active on exposure to light. On exposure to light the ferridoxin accepts electron (from splitting of water) and is reduced and thus becomes active. Then an enzyme ferridoxin-thioredoxin reductase reduces disulphide bond (-S-S) of thioredoxin to sulphhydryl (-SH-SH-) group. This reduced thioredoxin reacts with an inactive photosynthetic enzyme, reduces it and makes it active.

Fe-S protein

It is present in photosystem-I of chloroplast and contains four iron atoms bounded to two sulphur atoms. It is the primary acceptor of electron from PSI and is thereby reduced from Fe^{3+} to Fe^{2+}. However, during cyclic photophosphorylation in P_{700} it is oxidized back to Fe^{3+} by losing an electron.

Field Capacity

It is related to the capillary water in the soil that is present in capillary pores held against the force of gravity by adhesion between water molecules and soil particles. The soil is said to be wet at its field capacity when no more capillary movement of water occurs. For practical purposes field capacity of a soil is considered to be the upper limit of soil water for plant growth. It represents the water content of soil after downward drainage of gravitational water. It is also referred to as capillary capacity.

First Law of Thermodynamics

It states that energy can neither be created nor destroyed and it also states that total sum of energy of universe remains constant. Energy can be changed from one form to another.

Flavones

These are a class of flavonoides generally found in herbaceous family and are ivory to deep yellow colour pigments. They may be colourless, crystalline aromatic ketones in leaves, stem and seed capsule of many primrose. The colourless flavones absorb UV-radiation and act as protector against long wavelength UV-radiation.

Flavonoids

These are polyphenolic compounds with 1s- C-atom, 2 benzene rings that are joined by a linear 3 carbon chain. Flavonoids are secondary metabolites that are synthesized through phenylproponoid metabolic pathway and are found in all parts of the plant but mostly in flowers. They are water soluble and their main function is pigmentation in flowers, protection from microbes and insects. They are anti-allergic and anti-inflammatory.

Flavoprotien

These are flavin containing proteins and act as dehydrogenation catalysts in biological reactions. They contain a nucleic acid derivative of riboflavin; the FAD or FMN. The flavin that is present as a cofactor in this protein has spectroscopic property which makes it useful in studying enzyme activity occurring at active site. It is also used in bioluminescence, photosynthesis, DNA repair and even in oxidative stress tolerance by removing free radicals.

Florigen Concept

According to this concept florigen is a hormone required for flowering. This hormone can be transported between two living tissues and can induce flowering in the non-induced plants through grafting. Flowering stimulus is transported from leaves to buds and in general florigen moves through assimilates (phloem). This concept also lays emphasis on inhibitors of flowering. A non-induced leaf can inhibit florigen transport in a graft union may be because of inhibiting translocation of assimilates. This concept stresses on both promoting and inhibiting substances involved in flowering.

Fluid Mosaic Model

This model is given by Singer and Nicolson and is one of the most accepted models to explain the membrane structure. According to this model the membrane bilipid layer is quasi-fluid and the proteins are either integral or peripheral. Integral proteins are either embedded in the lipid bilayer or penetrate to its both sides and are tightly linked to it, while peripheral proteins are loosely attached at the outer surface of membrane. The fluidity of lipid bilayer offers free movement to proteins in membrane. Furthermore, the peripheral proteins may be attached either to polysaccharide forming glycoprotein or the lipid binds to carbohydrate forming lipopolysaccharide. Such an attachment helps in cell to cell recognition.

Fluorescence

It is observed when an excited electron returns back to its ground state. In this process light is emitted. Chlorophyll florescence is a well observed phenomenon in which deep red light is produced when an excited electron loses energy and comes to the ground state.

Fusicoccin

It is an organic compound that is produced by the fungus *Fusicoccum amygdale*. This compound acidifies the cell wall, and due to acidification the stoma opens irreversibly, water loss occurs and plants die.

G

Gallo Tannins

These are tannins obtained from tannic acid. They contain dimer or polymer of gallic acid that are joined to each other and also to sugar present in them. Gallo tannin is used commercially to tan leather as it denatures proteins by cross-linking them and prevents them from attack by bacteria. They are important to plants that produce them because they act as allelopathic agents. Many gallo tannins inhibit plant growth.

Gamma Plantlets

These are plantlets produced when seedlings are exposed to gamma rays emitted from cobalt-60 source. Such irradiation stops DNA synthesis and cell division which are necessary for seedling germination but still the seedlings germinate. Such seedlings die after a few weeks because in the absence of any cell division the seedlings cannot grow further.

Geotropism

In response to the gravity of earth roots grow downwards. This phenomenon is geotropism or gravitropism.

Gibberellic Acid

A number of gibberellins have been reported from both fungi and higher plants and they are denoted as GA_1, GA_2,

GA_3,......, and so on. Together called the gibberellic acid but since GA_3 is the first gibberellin to be reported and is the most studied, it is preferably called gibberellic acid.

Gibberellins

These are plant growth hormones which help in plant growth by enhancing stem elongation, leaf expansion. They are used to break dormancy of seed by working as substitute for low temperature and long day. They cause flowering and are used to overcome dwarfism. Gibberellins induce the production of enzymes like amylases, proteases, lipases and ribonuclease for starch mobilization during seed germination and seedling growth.

Global Warming

The increase in earth's average temperature due to increase in green house gases is global warming.

Glucosides

Glycosides are derived from glucose and give glucose on hydrolysis. The simplest glucosides are alkyl ethers.

Glutamine Synthesis

Glutamine is an important plant amide formed when an amino group is added to glutamic acid in the presence of enzyme glutamine synthetase coupling energy from ATP.

Glutathione

It is a tripeptide containing glutamate-cysteine-glycine. In it reduced form, it accepts sulfur atom from APS and is oxidized to GSSG. It acts as a reductant that helps in the protection of thio of enzymes. Glutathione acts as scavanger of free radical and thus protects plants from oxidative stress.

Glycolysis

It occurs in the cytosol of cell and in a series of reactions glucose is converted to two molecules of pyruvic acid.

Glycolysis occurs under anaerobic conditions and it is followed by Kreb's cycle.

Glycophytes

These are 'sugar loving plants' because they cannot grow in salty habitat. Plants generally are glycophytes. In salty soil their growth is decreased.

Glycosylation

It is the attachment of sugar to an enzyme or protein. Glycoprotiens are formed through glycosylation of glucose to peripheral protein in the membrane.

Glyoxysomes

These are spherical organelles bounded by a single membrane and are called microbodies. Glyoxysomes have granular cytoplasm and their main function is the conversion of fat to carbohydrate during seed germination. Glyoxsomes interact with oleosome in the conversion of fat to fatty acid and glycerol that enters glyoxysome. Here through β-oxidation they are converted to NADPH and acetyl CoA. These products form carbohydrate through glyoxylate cycle.

Grana

It is the structure associated with chloroplast. The internal membrane of chloroplast extended in the form of a flattened sac called thylakoid. These thylakoids are stacked together to form grana. In the grana the light reaction of photosynthesis occurs.

Gravitational Water

It refers to the water that percolates down in soil under the influence of gravity and reaches the water table. This type of water is not used by plants as it is not available to the roots of plant.

Gravitropism

It is the effect of gravity on the growth movement of root and shoot. Shoots show negative gravitropism, while roots show positive gravitropism. Root cap has statolith which acts as sensor and is therefore vital for gravity response. Plant hormone auxin also affects gravitropism.

Green House Effect

The phenomenon in which short wavelength solar radiation reaches the earth's surface by passing through CO_2 of the atmosphere, while the long wavelength radiation (heat) into which it is converted is largely prevented from leaving by the same gas is green house effect. This results in increase in the average temperature of the earth's surface.

Guard Cells

Present in leaves, these surround the stoma. Guard cells are in fact specialized epidermal cells that are present around stoma and contain chloroplast. The guard cells are kidney or bean shaped with thick inner wall. They help in the process of photosynthesis.

Guttation

It is the exudation of water in the form of droplets from the tip of leaves through specialized structures known as hydathodes. It generally occurs during warm humid nights. The water drops contain mineral salts, organic acids, carbohydrate and nitrogenous compounds.

H

Halophiles

The organisms which grow on hypersaline environments are halophiles. Many enzymes isolated from such organisms have a high requirement for salt concentration for stability and activity. Many of these organisms have hydrogen peroxide scavenging enzymes.

Halophytes

These are salt loving plants that grow in salty environment. Halophytes may be obligatory growing only under saline conditions or they may be facultative growing both in salty environment as well as without salt.

Harvest Index

It is the ratio of grain dry weight to total plants dry weight or yield biomass to total biomass at the time of harvest. Greater is the harvest index better is the allocation of carbon for grain production. It reflects the manner in which assimilates of photosynthesis are distributed between grain and vegetative parts of plant.

Heme Proteins

Iron containing proteins are called heme proteins, e.g., hemoglobin or the Fe-S proteins involved in photosynthesis.

Herbicides

These are killers of herbs or plants. Auxins like 2,4-D, NAA, 2,4,5-T are used as herbicides. An efficient herbicide is of low cost and kills dicot more than monocots. Herbicides are sprayed on plants and they enter leaves after absorption through cuticle and are translocated by the phloem. Herbicides are thought to alter DNA transcription and RNA translation thereby altering enzyme activity.

Herbivory

It deals with the protection mechanism present in plant against herbivores (plant-eating animals). The protection may be structural or biochemical. Many plants secrete substances which act as repellent to herbivores or may be toxic to them.

Heteroblasty

It is a unique feature found in desert plants to increase their rate of survival. It refers to the production of two types of seeds from the same plant. Such seeds differ morphologically and physiologically and have different requirements for germination. This helps the plant to adapt to xerophytic conditions because only some seeds germinate at any given time.

Heterophylly

It refers to the existence of two different types of leaves on the same plant. The primary leaves are generally simple while older leaves are compound. Perennial plants like junipers have needle-like juvenile leaves and scale-like adult leaves.

Heteropolymer

In this, different polymers are joined to each other in contrast to homopolymer in which similar polymers are joined. In a heteropolymer each polymer is coded by different gene. Therefore, in the formation of heteroploymer more than one gene is required.

High Irradiance Reaction

Light-dependent responses are known as high irradiance responses. Such reactions require light with energy and for a longer period. The magnitude of response by such reaction depends upon the energy rate and the time for which it is applied.

Hill Reaction

It was demonstrated by Robin Hill and R. Scarisbrick to confirm that O_2 is evolved from light driven splitting of water and not from CO_2. As per this reaction isolated chloroplast released O_2 when illuminated by light and a suitable electron acceptor was present to accept electron released by splitting of water. Thus some part of photosynthesis can occur in chloroplast without involvement of whole cell.

Homeostasis

It is the response of an organism to the variation in external environment conditions. The organism does not change its internal condition or changes it to certain limit even if the external environment in which it lives changes drastically. Such an adaptation to the changing environment without undergoing any change in itself is known as homeostasis.

Hormones

These are organic substances that affect the physiological processes even at low concentration. They occur naturally in both plants and animals. In plants they are referred to as phytohormones.

Horse Latitude

This latitude extends to 30° north and south of equator. In this region precipitation is less and therefore it is a desert area.

Horticulture

It is the science that deals with vegetables or ornamental crops.

Humus

It is dead decomposed soil organic matter that is dark, finely particulate and highly amorphous. It consists of C, H, O and N. During humus formation plant materials are decomposed to such an extent that they are not recognizable.

Hydration

It is the process of adsorption of water to any surface. Water molecules are polar in nature and are bound by hydrogen bonds to other polar molecules. It is through hydration that seeds adsorb water and then only they can germinate.

Hydrogen Bond

It is formed between H-atom of water molecules and an electronegative atom like O and N.

Hydrophilic

Such molecules are water friendly and can form bond with water through hydrophilic bonding.

Hydrophobic

These are water hating molecules and avoid contact with water. They combine with other hydrophobic molecules through hydrophobic bond as in biological membranes where the non-polar hydrophobic groups are responsible for the formation of lipid bilayer as they form bond among themselves avoiding contact with water.

Hydroponics

It is soil-less culture in which all nutrients are added in the liquid medium. The nutrients are added in a known amount and the plants grow by absorbing nutrients from solution.

Its roots freely float in solution, but the solution must be aerated to avoid anoxia (that inhibits root growth and reduces the uptake of nutrients). In hydroponic culture the nutrients should be added at some interval regularly to avoid ion depletion and change in pH as roots take up nutrients.

Hypocotyls

It is the portion of embryonic axis just below the cotyledon. During germination in seeds the hypocotyl forms root.

Hypogeal Germination

In this kind of germination the seeds elongate in the epicotyls, i.e., between cotyledons and shoot apical meristem. Shoot apical meristem forms the shoot that grows above ground while the cotyledon remains below the ground. The shoot grows in a hook shape and pulls the shoot tip above the soil. Such germination is called hypogeal because cotyledons remain below the ground.

Hyponasty

It is a nastic movement in which differential growth occurs. In this the petiole and leaves are pointed upward. It may occur due to gibberellins. The unequal flows of hormones are responsible for such nastic movement.

Hypoxic

It refers to reduced oxygen concentration. In an aquatic system when eutrophication occurs then the concentration of dissolved oxygen (DO) is lowered and this leads to hypoxic condition in the aquatic system which causes death of fish. Generally DO in the range of 10-30% saturation is called hypoxic.

I

Ideotype

A plant with its morphology ideally suited for getting maximum produce under a particular condition.

Imbibition

It is a surface phenomenon in which water moves into substances such as wood, seed, agar and causes an increase in their volume. For imbibition to occur the substances that imbibe (adsorbent) must have affinity for the substance that is imbibed and a water potential gradient must exist between them. Imbibition is very important in plants because it is through this process that water enters into roots when adsorption of water occurs by the cell wall of root hairs. Seed germination occurs after imbibing water.

Impaction

It is a process by which dormancy in seed is broken by shaking the seeds. On shaking the cork-like filling (strophiolar plug) is removed from the seed coat. It is this plug that interferes with water and oxygen entry into seeds causing dormancy. Removal of this plug allows the seeds to germinate.

Incipient Plasmolysis

The stage at which protoplasm just shrinks from the membrane and gets separated from the cell wall is incipient plasmolysis.

Induced Fit Hypothesis

According to this hypothesis when a substance comes in close proximity of an enzyme's active site it changes the conformation of enzyme in such a way so as to form a functional enzyme-substrate complex. For an active enzyme-substrate complex either the enzyme or the substrate must undergo conformational change.

Inhibitors

These are foreign substances that block the activity of enzymes by combining with the enzyme. They may be competitive inhibitors if they are structurally similar to substrate and combine with the enzyme active site preventing substrate binding to enzyme or they may be non-competitive inhibitors which combine at site other than active site but they change the conformation of active site in such a way that substrate fails to bind to it. Both types inhibit the enzyme activity.

Integral Proteins

These are internal proteins of the membrane. They are embedded in the lipid bilayers. Their hydrophobic part penetrates the lipid bilayer and sometimes extends out of the lipid bilayer. Such proteins cannot be easily removed unless all H-bonds present in the membrane are broken.

Ion

It is a charged particle formed when an atom or molecule loses or gains valence electron. When it loses electron it forms cation (positively charged ion) and when it gains electron it forms anion (negatively charged ion).

Ionization

It is the process of converting an atom or molecule into an ion by adding or removing electron.

Ionophore

These are small molecules that help in the transport of ions across the lipid bilayer.

Iodine Number

The number of halogens absorbed by 100 g of oil is referred to as iodine number of oil.

Ionic Bond

It is formed when an atom loses electron and gives it to another atom that accepts this electron. The atom that loses electron is positively charged while the receiving atom is negatively charged and the two are held together by the force of attraction between two differently charged particles.

Irradiance

It is the flux of light energy on a surface. Infact light plays a very important role in plant development and is essential for photosynthesis. The range of light 400-700 nm is called PAR and is used in photosynthetic process.

Isozymes

These are different forms of the same enzyme which catalyse the same reaction and form the same product. Their difference lies in their genetic make-up.

J

Jasmonic Acid

It is a growth regulator found in several plants. It causes senescence in detached leaf segment, but it also provides resistance to insects and diseases. Therefore, it accumulates in wounded plants and also in plants treated with elicitors. It is synthesized from linolenic acid. Besides providing resistance, it also has various physiological functions. It induces seed germination, root development, tendril coiling etc. It generally works in combination with ethylene.

Juvenility

The difference in morphology of a juvenile leaf from adult leaf is juvenility. During differentiation a plant passes through juvenility.

K

Kinetic Energy

It is the energy that a body possesses by virtue of its motion. It depends upon mass and velocity.

Kinetin

It was obtained from degraded autoclaved DNA. It is an adenine like active substance that is involved in cytokinesis or cell division. Kinetin is 6-furfurylamino purine. Kinetin itself is not found in plants it is cytokinin that is present in plants.

Kranz Anatomy

Kranz anatomy is present in the leaves of C4 plants. In this type of leaf anatomy the leaf has two types of photosynthetic cells; mesophyll cell and bundle sheath cell. The bundle sheath cell has chloroplast, although larger in size they have less developed grana and store starch. These bundle sheath cells are present around vascular bundles and separate them from mesophyll cells. This concentric arrangement of bundle sheath cells is called the Kranz anatomy.

Kreb's Cycle

Also known as citric acid cycle or tricarboxylic acid cycle (TCA). It occurs in mitochondria during respiration. The product of glycolysis, i.e., pyruvate undergoes breakdown in mitochondria to form acetyl Co-A which reacts with oxaloacetic acid (OAA) to form citric acid and the cycle starts.

L

Latent Heat of Condensation

Latent heat is the energy change that occurs when one state transforms into another. It is the invisible heat change accompanying the state change process. The latent heat of condensation is the amount of heat energy released when a gas changes its state to a liquid.

Latent Heat of Vaporization

It is the amount of heat energy that is required to change a liquid to gaseous state.

Law of Minimum

According to this law the growth of a plant depends upon the amount of factor that is present in a limited amount or in least amount. Even if all factors are sufficient still growth will not occur unless the limiting factor is provided.

Law of Tolerance

This law states that there is a limit for any factor to which the plant responds. The plant will not respond if the factor is present at some minimum level but if it is present at some maximum level then it will be toxic to plant. Thus plant responds at an optimum level of the factor. The range between minimum and maximum level is the tolerance range.

Leaching

It is the removal of materials by dissolving in water. It is through this process that water removes and transports soil humus and inorganic nutrients.

Leaf Area Index

It refers to the ratio of leaf surface area to the ground area that is occupied by that crop. Leaf area measurement is an important parameter as leaves are primary photosynthetic organs of plant and sometime growth is expressed on a leaf area basis.

Leaf Area Ratio

Ratio of leaf area per plant to dry weight is referred to as leaf area ratio.

Leaf Mosaic

The leaf has a mosaic pattern, i.e., light and dark coloration can be seen on the leaf. In leaf mosaic irregular yellowing occurs in the form of patches.

Leaf Primordium

The newly formed leaf arising from shoot apical meristem is the leaf primordium.

Leghaemoglobin

It is a pigment very similar to haemoglobin of blood and gives pink colour to the healthy legume nodules. Leghaemoglobin is synthesized symbiotically, its globin part is synthesized by plant whereas heme group by *Rhizobium* (a nitrogen fixing bacteria). Leghaemoglobin has high affinity for oxygen. It is this feature that is very important in nitrogen fixation because the nitrogenase complex (enzyme involved in nitrogen fixation) are very sensitive to oxygen and leghaemoglobin binds oxygen maintaining its level low thus preventing damage.

Lenticles

These are structures involved in gaseous exchange. These are found in the periderm (bark) of most plants as a small area of loosely arranged thin walled cells with many intercellular spaces. Lenticels are also called complementary tissues or filling tissues and the pressure exerted in their formation ruptures the epidermis to expose the lenticles.

Light Compensation Point

It is the light level at which CO_2 output through respiration is equal to the CO_2 input through photosynthesis. The two processes are balanced in terms of CO_2 at this light level.

Light Harvesting Complex

This complex collects light energy and transfers this energy to the reaction centres for light dependent reaction. This complex consists of photosynthetic pigments like chlorophylls and carotenoids which surround the photosynthetic reaction centre, absorb light energy and pass it to the reaction centre through resonance.

Light Line

A continuous line through the epidermis seen in section of certain legumes is light line.

Lignin

It is a chemical present in the secondary cell wall and adds rigidity to the wall structure. It is the second most abundant organic compound on earth, only after cellulose. Lignin is mostly found in large amount in wood where it accumulates in middle lamella, primary and secondary wall of xylem elements. Besides providing strength lignin also provides protection against attack by pathogen.

Limiting Factor

Any factor that is present in least amount compared to other factors serves as a limiting factor and retards the rate of a process. If this factor is supplied the rate of process increases unless some other factor becomes limiting. This factor controls a process.

Limnetic Zone

It is also known as a zone of swimming organisms. This is the zone of open water down to the depth of light penetration. It contains phytoplanktons.

Linolenic Acid

It is an essential fatty acid that is required by the body and is synthesized in plants. It is of two types; Alpha-linolenic acid found in many vegetable oils and Gamma-linolenic acid.

Lipases

These are water soluble enzymes that break the ester bond present in lipids. Lipases perform essential role in living organisms as they help in the digestion of lipids.

Lipids

These are fat soluble (hydrophobic) molecules that perform various important functions. They are components of cell membrane as lipid bilayer, they store energy and also participate in signalling. Lipids include fat, oil, wax, sterol, cholesterol, fatty acids, phospholipids etc.

Lipopolysaccharide

It has both lipid and polysaccharide joined by covalent bond. It is an important constituent of bacterial outer cell membrane where it helps in cell recognition and also provides protection from attack by chemicals.

Littoral Zone

This is the zone from shore line to inner most rooted plants.

Logarithmic Phase

It is the phase of exponential growth. This phase is followed by stationary phase, where growth stops and finally reaches the death phase. Maximum growth occurs in the log phase.

Long Day Plants

These plants require long day and short night to flower. It was later found that it is uninterrupted dark period that is critical for flowering. If the dark period is interrupted by brief flashes of light flowering fails even if day length is longer than critical value. Examples of long day plant are spinach, barley, radish, carrot etc.

Long-Short-Day Plant

Such plants require a long photoperiod followed by short photoperiod to flower, e.g., *Bryophyllum.*

Luxury Consumption

In adequate zone, accumulation of nutrients occurs in vacuoles. This additional accumulation is luxury consumption.

M

Macronutrients

These are essential nutrients required in the concentration of 1000 μg g^{-1} of dry matter or more for normal growth and development of plants. These include C, H, O, P, K, N, S, Ca, Mg. These are also known as major elements.

Matric Potential

It is the amount of work done to move pure water present around soil into soil at atmospheric pressure. The water inside and outside are at same elevation, so there is no effect of gravity on this potential or it is the energy lost as water diffuses and interacts with other substance in the medium. It is the component of the water potential resulted due to the interaction of the water with colloids and to capillary forces.

Megapascals

Force of one newton per square meter is one megapascal. (10 bars = 1 Megapascal)

Melavonic Acid Pathway

It is an important cellular metabolic pathway present in all higher eukaryotes and in many bacteria. This pathway is a part of terpenes and steroid biosynthesis. It is also used in the production of dimethylallyl pyrophosphate and isopentenyl pyrophosphate.

Meniscus

It is a curve on the surface of a liquid. It may be concave or convex. Formation of meniscus depends upon the attraction between the surface of container in which it is present and the liquid itself. If they have greater attraction to each other then convex meniscus is formed, e.g., mercury and glass, else concave meniscus is formed, e.g., water and glass.

Mercaptans

These are sulphur containing organic chemical substances with the general formula R-SH where 'R' may be any group. They are used as chain transfer agents in polymerization and are also used as odourizing agent in natural gas. They are produced naturally in human body during digestion of garlic or some other food.

Meristem

It is a tissue made up of undifferentiated cells and is required to produce new cells. It is found in all plants wherever growth can occur. Meristamatic cells divide to produce a new cell that helps in expansion and differentiation of tissues. They form the primary structure of plant body.

Mesophyll Conductance

The ease with which CO_2 diffuses through the mesophyll cells is known as mesophyll conductance.

Mesophyll Resistance

The resistance offered by the mesophyll cells for diffusion of CO_2.

Mesophyte

These are terrestrial plants that grow in moderate amount of moisture. They have well developed root system and vascular tissue. They usually have larger and thinner leaves in comparison to xerophytes. Mesophytes lack xerophytic adaptation and thus they cannot survive under drought conditions.

Micelles

These are formed when a substance having polar head and non-polar tail is added to water. The non-polar tail of the molecules comes in the centre of micelles while polar head is outside for interaction with water molecule.

Micronutrients

These are also known as trace elements. These are essential nutrients required in the concentration of less than 1000 $\mu g\ g^{-1}$ of dry matter. They have some catalytic and regulatory roles in plants, e.g., Cl, B, Fe, Mo, Zn, Cu, Ni and Mo.

Metabolism

It is the sum of all processes that occurs in a living organism in order to maintain life like growth, reproduction or even response to external factors. Metabolism may be catabolic if large molecules are broken down or anabolic if large molecules are synthesized from small molecules.

Mineral Nutrition

Except for carbon, hydrogen and oxygen other nutrients required by plants are obtained from soil and the source of these nutrients are minerals. Thus the nutrition of plants is said to be mineral nutrition.

Mitochondria

It is known as the 'power house of cell' because ATP synthesis occurs here. It is a semi-autonomous organelle as it can divide and carry out its functions.

Mitosis

It is an equational cell division in which parent cells give rise to two daughter cells with equal number of chromosome as the parent cell. It occurs in all cells to maintain growth.

Molality

It is the number of moles of solute that are dissolved in 1 kg of solvent. It is independent of temperature.

Molar Extinction Coefficient

It is the measurement of the absorbance of a chemical at a given wavelength.

Molarity

It is the number of moles of solute that are dissolved in 1 liter of the solvent. It depends upon temperature.

Mole Fraction

It is the ratio of number of moles of a solute to the total number of moles of solute and solvent.

Monocarpic Species

These species flower only once in their lifetime and die after flowering.

Monoecious

These plants have male and female flowers on single stem although at different positions, e.g., maize, pumpkin etc.

Mucigel

It is secreted by the root cap and is a polysaccharide rich slime. It prevents root desiccation and helps in its easy penetration into the soil. Mucigel also helps in diffusion (uptake) of water and mineral nutrients from dry soil into roots.

Mycorrhiza

It is a symbiotic association of fungi with roots of plants. The fungi get organic nutrients from plant and in turn it helps in improving the water absorbing capacity of root and also mineral uptake. In such association root epidermal and cortical cells are involved. Mycorrhiza are of two types; ectomycorrhiza and endomycorrhiza.

N

NADPH

It is reduced pyridine nucleotide and acts as an electron donor. It is produced during light reaction of photosynthesis and also in Kreb's cycle and for each molecule of NADPH three molecules of ATP are synthesized in the electron transport chain.

Nastic Movement

It is the non-directional response of a plant to any stimulus. The stimulus may be light (photonasty).

Necrosis

It is a disease of leaf in which leaf has dead tissues.

Nerst Equation

It is used to analyze whether transport of an ion inside the cell is an active process or a passive process. It is given by the equation.

$$E = RT/nF \ln [Co^+]/[Ci^-] \text{ or } [Ai^-]/[Ao^+]$$

Where,
E = Transmembrane potential
R = Gas constant
T = Absolute temperature
n = Valence of ion
F = Faraday's constant

Ao^+ and Co^+ = Anion and cation concentration outside the cell
Ai^- and Ci^- = Anion and cation concentration inside the cell

If E measured is greater than E calculated then for cations, active transport of ions occur and for anions passive transport of ions occur. Similarly, if E measured is less than E calculated then for anion active transport occurs and for cation passive transport occurs.

Neutralization

It occurs when an acid reacts with a base in equal amount and in aqueous solution. The H^+ ion of an acid in solution combines with the OH^- ion of base to form water and the other ions of acid and base form salt and do not take part in the reaction.

Nitrate Reduction

Plants absorb nitrogen from the soil in the form of nitrate. But this nitrate cannot be used as such by plant and is reduced to ammonia. First of all the nitrate is reduced to nitrite by nitrate reductase. In this process two electrons are accepted by nitrate and nitrite is formed. This nitrite is then reduced to ammonia by accepting electrons in the presence of enzyme nitrite reductase and energy is taken from the process of respiration.

Nitrification

It is the process of converting ammonia into nitrate. Nitrification occupies a central position in the nitrogen cycle. It increases the nitrogen supply to plants.

Nitrogen Fixation

It is the process by which nitrogen present in the atmosphere in its inert molecular form (N_2) is converted into nitrogenous compound like NH_3, NO_3 and NO_2. Nitrogen fixation may occur biologically or non-biologically. Biological nitrogen fixation is performed by cyanobacteria, actinobacteria and certain anaerobic bacteria. *Rhizobium* present in the root nodules of leguminous plants is well known for nitrogen fixation. Non-biological nitrogen fixation can occur by the process such as lightning, combustion etc.

Nitrogenase

It is an enzyme used to fix atmospheric nitrogen. It catalyzes the reduction of nitrogen into ammonia.

Nodules

Nitrogen-fixing association between the microsymbiont (bacteria) and the host (generally a leguminous plant) results in the development of enlarged multicellular structures known as nodules, found on the root of the host plants.

Non-Competitive Inhibition

This is a kind of inhibition in which the inhibitor of an enzyme is structurally different from substrate and bind at sites other than the active site. Their effect cannot be overcome by increasing substrate concentration because they do not compete with substrate. They change the enzyme structure by either combining with or destroying the structure of sulfhydryl group, e.g., such inhibitors are toxic metal ions.

Non-Cyclic Photophosphorylation

It is the non-cyclic process for formation of ATP during photosynthesis. The electrons released from PS II by the splitting of water are accepted by various intermediates before being accepted by PS I. From PS I the electron finally reaches NADP but this electron is never cycled back to PS II. This overall process requires energy that is taken in the form of light.

Non-Reducing Sugars

These sugars do not have free aldehyde or ketone groups and therefore do not have reducing property, e.g., sucrose.

Nucleotide

It is an organic compound that is formed when a nitrogen base is added to sugar and a phosphate group. Bases are of two types; purines and pyrimidines. The sugar may be ribose or deoxyribose. Nucleotides are the building blocks of RNA and DNA and they also act as cofactors in various metabolic processes.

Nutrient Solution

The solution having all essential mineral salts needed for plant growth.

Nyctinasty

It is a nastic movement performed by higher plants as darkness approaches. Nyctinasty is a circadian rhythm and is dependent on diurnal light and temperature changes. The example of this movement is the sleep movement in the leaves of many legumes or the closing of the petals of flower as darkness approaches.

O

Obligate Halophytes

These are true halophytes that cannot grow in the absence of salt. They need salt for growth because in non-salty soil they cannot compete with glycophytes. Sometimes they accumulate large amount of salt in their cell and therefore cannot survive in non-salty soil.

Operational Environment

It is the environment in which the plant lives and that is important for plant survival. It determines the structural features of plant and includes climatic, soil and biotic factors.

Organelle

It is a part of the cell. It is a separate subunit, surrounded by its membrane and performs a specialized function in the cell.

Organic Acid

These are organic compounds with acidic property. The acid character is because of the presence of carboxyl (-COOH-) group or because of H^+ ion released in the solution.

Organogenesis

It is the genesis (formation) of an organ from an unorganized mass of cell (callus). The callus gives rise to shoot and adventitious root. For organogenesis cytokinin-auxin ratio is important; higher ratio of cytokinin-auxin gives rise to shoot. The organogenesis may be direct or indirect.

Orthogravitropism

It is the movement of phytohormones or mobile particles under the influence of gravity. The vertically oriented stem and root axis also show orthogravitropism.

Osmolality

It is the total number of molalities of all solutes in a solution. It can be defined as the count in moles of the osmolality active particles in a solution and this is independent of the size of the particles.

Osmometer

It is an instrument used to measure osmosis in laboratory. A living cell itself is an osmotic system where various solutions are separated from each other by membrane and a pressure is built up in any one solution at any time. The cell wall rigidity is responsible for increase in pressure.

Osmoprotectants

Osmotic solutes like proline, glycinebetaine etc. which are suggested to be protectants of certain enzymes are known as osmoprotectants.

Osmoregulation

This could be done by increasing the level of proline, betaine, carbohydrates etc. It is observed that the level of salt in a cell is kept low by storing them in vacuole or by diluting their concentration through osmoregulation under salt stress. It is defined as regulation of osmotic pressure resulting in a constant internal osmotic pressure while the external osmotic pressure is varying e.g., in certain algae.

Osmosis

It is a process in living cell in which solvent moves from a region of higher water potential to the region of lower water potential across a semi-permeable membrane. When a cell is placed in hypertonic solution water comes out of the cell by the process of exosmosis while in a hypotonic solution water enters the cell by endosmosis.

Osmotic Adjustment

It is accomplished by accumulation of compatible soluble solutes. Osmotic adjustment is a net increase in solute content for cell and should not be confused with the increase in solute concentration resulting due to cell dehydration. This results in the maintenance of turgor at lower water potential. These compatible solutes are mainly organic compounds, e.g., proline and glycinebetaine and sometimes sugar alcohols. In certain cases there can be inorganic contents like potassium. An additional expenditure of energy is required for enhanced ion uptake (and transport) and synthesis of organic solutes in the adjustment to osmotic stress. As a consequence actual yield levels in genotypes which exhibit osmotic adjustment may be lower under stress conditions.

Osmotic Potential

It is also called solute potential and is represented by ψs. In osmosis water moves from higher water potential to lower water potential through a semi-permeable membrane. Presence of solute in a solution lowers the water potential. Osmotic potential refers to the potential of a solution to allow flow of water into it as a result of dissolved solutes under an ideal osmotic condition in the absence of energy. It always has negative value and is measured in kilo pascal (KPa).

Osmoticum

Organic compounds at times added to nutrient solution to lower the water potential, e.g., polyethylene glycol.

Oxidation Reduction

It is the change in oxidation number without any actual transfer of electrons. When oxidation number is increased oxidation occurs and when oxidation number is decreased reduction occurs.

Oxidative Phosphorylation

It occurs in the mitochondria and here ATP is produced by causing reduction of oxygen. During oxidative phosphorylation in mitochondria electrons move finally from NADH to O_2.

Ozone

Ozone absorbs energy of ultraviolet radiation. Its depletion could cause carcinoma and severe phytotoxic diseases in living beings.

P

Parahelionastic

The leaf movement resulting into reduced incident radiation on leaf is parahelionastic.

Parthenocarpic Fruit Development

Fruits develop from ovary without fertilization and they lack seeds. The fruit develops when pollination does not occur as in citrus and banana. It may occur when pollination occurs but fertilization fails as in some orchids and it may also occur due to zygote sterility, i.e. zygote is formed but either it aborts or is sterile as in grapes and cherries.

Parthenocarpy

It is the phenomenon of development of fruit from ovary without fertilization. Such fruits lack seeds and are called parthenocarpic fruit. These fruits have more pulp than normal fruits. Pathenocarpy occurs when pollination is not followed by fertilization or if embryo is formed but aborts, e.g., banana, citrus, grape, pineapple, varieties of apple and pear.

Passive Transport

It does not require energy for transport of molecules across a membrane. This type of biological transport occurs in the direction of concentration gradient, i.e., from higher to lower concentration and hence no energy is needed for transport.

Pasteur Effect

When a tissue is shifted from anaerobic to aerobic condition, fermentation is suppressed and rate of consumption of respiratory substrate is reduced. This phenomenon is Pasteur Effect. It refers to the inhibition of carbohydrate breakdown due to the presence of oxygen. Oxygen acts as an allosteric inhibitor for phosphofructokinase and decreases glycolysis and fermentation. Pasteur effect has practical utility in preventing fruit ripening by decreasing O_2 to a concentration where aerobic respiration is minimum but the breakdown of sugar anaerobically is also not stimulated.

Pentose Phosphate Pathway

In plants besides glycolysis and Kreb's cycle another pathway that is used for oxidation of sugar and production of ATP is the pentose phosphate pathway (PPP) or the Hexose monophosphate shunt or phosphogluconate pathway. It is called pentose phosphate because 5 carbon sugar phosphates are its intermediates. PPP occurs in the cytosol and their intermediates are similar to those of glycolysis and Calvin cycle. The reaction starts with glucose-6-phosphate and at the end various pentose phosphates are formed. In the process, 3-phosphoglyceraldehyde and fructose-6-phosphate are also produced which are intermediates of glycolysis and therefore it is an alternate route for the breakdown products of glycolysis. PPP also gives $NADPH_2$ that is used for production of ATP, erythrose-4-phosphate that is used for production of phenolic compounds and ribose-5-phosphate used for formation of RNA and DNA through formation of deoxyribose and ribose sugar.

Peptide Bonds

These are present in the structure of protein and join two amino acids. The α-amino group of one amino acid is joined to the α-carboxylic group of another amino acid by the peptide bond $\overset{\displaystyle O}{\overset{\|}{C}}-\underset{\displaystyle H}{\underset{|}{N}}$ and water is removed.

Peribacteroid Membrane

In the root nodules there is presence of group of enlarged non-motile bacteria called bacteroids. These bacteroids occur in groups and are surrounded by peribacteroid membrane.

Pericycle

In the internal structure of root and shoot pericycle is visible outside the vascular tissue inner to the endodermis. It is composed of sclerenchymatous cells that are either in the form of patches or may form complete cylinder. Pericycle is the site of origin of lateral roots and the entire phellogen originates from it.

Peripheral Proteins

These are present in the membrane at its outer surface and can be easily removed.

Permanent Wilting Point

It is also referred to as permanent wilting percentage. It is the soil water content at which plant remains wilted unless water is added to the soil, usually when the soil water potential reaches -1.5 MPa. Most plants cannot obtain sufficient amount of water at soil water potential below -1.5 MPa.

Permeability

It refers to the property of membrane to allow certain substances to pass through it and restrict others. The substance passing through the membrane must be in dissolved form in water. The membrane in general is only differentially permeable allowing only some substances to pass through it and not all the substances.

Permeases

These are involved in transport of proteins and are present in the biological membrane. They are transmembrane proteins and help in the diffusion of specific molecules in or out of the cell.

Phase

In a free running cycle of a biological clock phase represents the most obvious point that can be easily identified by its relationship to the rest of the cycle. For example, maximum and minimum are phase points. However, the maximum or the peak is the most important reference point for phase relationships and is referred to as phase.

Phellem

Phellogen gives rise to phellem towards the outer side by periclinal divisions. Phellem is also known as cork and its cells are dead at maturity. Phellem cells are arranged in radial rows and are suberised.

Phelloderm

Phellogen undergoes periclinal divisions to form phelloderm towards the inner side. Pheloderm is also known as secondary cortex and its cells are living having simple pits on their wall. They are arranged radially and sometimes may contain chloroplasts in some plants.

Phellogen

It is also called cork cambium and is a secondary lateral meristem. It is a part of periderm that is formed when epidermis ruptures due to secondary growth in extractable region. In fact secondary xylem exerts pressure on cells outside it and epidermis ruptures and is replaced by protective periderm. The cells of phollogen are only of one type and are highly vacuolated. Phellem undergoes periclinal divisions to form phellem towards outer side and phelloderm towards inside.

Pheophytin

It is formed when magnesium ion is lost from the chlorophyll molecules. It is non-green product and in some amount it may occur in the chloroplast as electron acceptor. It transfers electrons to the first acceptor in PSII.

Pheromones

These are chemical substances that trigger behavioral or physiological responses in another member of the same species. This is common in animals but plants also release pheromones.

Phloem

Phloem is a complex vascular tissue that functions as food transporter. It includes sieve elements, companion cells, phloem parenchyma and phloem fibres. The sieve elements of phloem consist of sieve cell, sieve tube and sieve plate and they are the conducting elements of phloem. The companion cells are associated with sieve tube elements and help in the maintenance of pressure gradient together with sieve tube. They also help in passage of materials. Phloem parenchyma cells are parenchymatous living cells in phloem concerned with storage and lateral transport of food material and water. The phloem fibres are elongated cells with lignified walls and simple pits. They provide mechanical support but also play a role in food transport.

Phloem Loading

It is the process of transport of assimilates from leaf mesophyll cells to phloem sieve-tube element. Such a transport requires energy because photosynthates in leaf mesophyll cells are at lower osmotic potential. Phloem loading is most easy in C4 species having vascular bundle sheath cells with chloroplast. In this the chloroplast provides energy in the form of ATP after illumination. During this process assimilates are transported from symplast of leaf mesophyll cells to cell wall apoplast and finally to of phloem sieve tube elements.

Phloem Unloading

It is the movement of assimilates from phloem sieve tube elements to the cells which need it or rather to the sink. It occurs either through symplast of phloem sieve tube to the symplast of sink cells or may occur via apoplast of phloem cell wall.

Phospholipids

These belong to a class of lipid and contain a diglyceride, a phosphate group and an organic molecule such as choline. They are structural components of the biological membrane and are present in the form of bilayer having a hydrophobic tail and hydrophilic head.

Phosphorescence

It is the light produced by some substances after absorbing energy. The phosphorescent substance absorbs energy and its electron gets excited, and when they fall back to their original energy level they release energy in the form of light. Phosphorescence is different from fluorescence because it emits light after radiation causing its emission is removed.

Photo Reactivation

It is a process of DNA repair by using visible light. During the formation of photodimer caused by UV light two pyrimidines are linked instead of a pyrimidine-purine pair. To repair the damage photolyase enzyme is used to break the bond between the two pyrimidines and to form the initial base pair. This enzyme works in the presence of visible light (300-600 nm) and this type of repair by using visible light is photo reactivation.

Photochemical Smog

In the presence of sunlight SO_2 and NO_2 combine to form more harmful pollutant known as photochemical smog.

Photolithotrophic

Green plants are photolithotrophic which means that essential components of growth are synthesized in the presence of light from inorganic or lithic (soil) elements.

Photomorphogenesis

The change in plant growth and development under the influence of light is photomorphogenesis. Pigments like phytochrome (absorbing red and far-red light) and cryptochrome (absorbing violet and blue light) are important for photomorophogenesis.

Photon

It is an elementary particle and is a carrier of electromagnetic radiation of all wavelengths. Photon has zero rest mass and it can travel in vacuum at the speed of light. A photon has both wave and particle property.

Photoperiodism

It is the response of plants to the relative length of day and night. Some plants require short day to flower and are called short-day plants while others require a long day to flower and are called long-day plants. However, some plants have no specific requirement of day length to flower and are day-neutral plants. Stimulus of critical length of dark period is important.

Photophosphorylation

It is the process of formation of ATP by using light energy. Light energy is used to excite electrons from pigment system which move to the electron transport chain and during the process ATP is produced.

Photorespiration

It refers to the respiration that occurs in the presence of light in photosynthesizing tissues. The functional enzyme is Rubisco. Rubisco acts both as carboxylase and as oxygenase. When concentration of O_2 is high Rubisco works as an oxygenase and photorespiration occurs. The first reaction of photorespiration is the synthesis of 3PGA and 2 phosphoglycolic acid in chloroplast. In subsequent reaction 2-phosphoglycolic acid is converted to glycolic acid that enters peroxysome where it is oxidized to glyoxylic acid which is then converted to glycine. Glycine enters mitochondria and is deaminated and decarboxylated to serine releasing NH_3 and CO_2. PGA enters into Calvin cycle in the chloroplast. Photorespiration although reduces photosynthesis still it is useful as the amino acid glycine and serine are precursors of important metabolites likes proteins, chlorophyll, nucleotides etc. It also helps in dissipating excess reducing power by consuming NADPH generated in light reaction during the conversion of glycolic acid to glyoxylic acid.

Photosynthesis

It is the process that occurs in the chloroplast of leaves and is responsible for synthesis of carbohydrates. In this oxidation-reduction process water is oxidized to O_2 and CO_2 is reduced to carbohydrate. Oxygen released as a byproduct is used in respiration and carbohydrate is the main source of energy for plants. In this process light of sun is used as the source of energy. The process involves two steps:

i) Light reaction that occurs in grana of chloroplast
ii) Dark reaction that occurs in stroma

The products of light reaction (NADPH and ATP) are used in dark reaction.

Photosynthetic Pigments

These are the pigments which are involved in photosynthesis. These mainly include chlorophylls and carotenoids. Chlorophylls are the green pigments in plants which absorb light and emit electron which pass through pigment systems I and II. Carotenoids transfer light energy to Chl a molecules and also protect the photo-oxidation of chlorophylls. In addition to these pigments anthocyanin and pheophytin are the other photosynthetic pigments.

Photosynthetic Unit

The number of pigment molecules required to effect a photochemical act of absorption and transfer of energy is known as photosynthetic unit or quantasome.

Photosynthetically Active Radiation (PAR)

This radiation is used by plants for photosynthesis and it lies in the spectral range of 400-700 nm. Below this range the shorter wavelength are very energetic and can destroy cells and tissues while above this range the wavelength has less energy and cannot carry out photosynthesis. PAR lies in the range of light visible to human being.

Photosystems

Photosystems are protein complexes involved in photosynthesis. Phtosystems are reaction centres providing the driving force for electron and proton transfer. They absorb light and release electrons. A series of oxidation-reduction reactions occur and finally the terminal acceptor is reduced.

Photosystem I (P_{700}) is most reactive at 700 nm and it is found in chloroplast and in green-sulphur bacteria. It emits electrons when excited by light which is accepted finally by ferridoxin.

Photosystem II (P_{680}) is most reactive at wavelength of 680 nm and is found in chloroplast or non-sulphur purple bacteria. It emits electrons which are accepted terminally by quinones. The splitting of water occurs in this photo system.

Phototropism

It is the light affected direction of growth. In the process light acts as a stimulus and growth occurs in the direction of light source. It is observed mainly in plants but fungi also show this movement. If growth occurs in the direction of light it is a positive phototropism e.g., stem. The growth occurring away from light is negative phototropism e.g., root. Phototropic response depends upon wavelength of light, for example, stem tip shows positive phototropism to blue-light, while root tip shows negative phototropism. Phototropism is enabled by auxin.

Phyllotaxis

It is the leaf arrangement present on stem. The leaves may be arranged in four basic patterns: alternate, opposite, spiral and whorled.

Phytin

It is a Ca-Mg salt of phytic acid derived from various seeds, potatoes etc. and is used as a dietary supplement of Ca and it also acts a phosphate depot.

Phytoalexins

These are substances (may be phenolic or terpenoid) produced by plants against attacking organisms. This phytochemical provides protection against attacking organisms either by puncturing cell wall, disrupting its metabolism or preventing its reproduction. They inhibit the growth of certain microorganisms.

Phytochrome

These are pigments that are used by plants to detect light. Phytochromes are sensitive to red and far-red light. They are functional in processes like photoperiodism, germination of seed, elongation of seedlings, chlorophyll synthesis or in making the hook of hypocotyl straighten in dicot.

Phytoferritin

It is a crystalline or paracrystalline body found in plastids of cambium in willow. It is found only when the plants are grown under dim-light conditions. These are not viral particles as they are unaffected by deoxyribonucleases and ribonucleases.

Pigment

It is a colouring substance in cells that absorb a particular wavelength of light. All types of cells have pigments. In plants pigments are grouped into four major classes.

1. Chlorophylls
2. Carotenoids
   i) carotenes
   ii) xanthophylls
3. Flavonoids
4. Betalains

Plagiotropic Movement

These are movements of plants based on the stimulus it perceives. The direction of the movement is oblique to the direction of the stimulus.

Plant Growth Regulators

These are organic substances that affect the physiological processes in plants even when present in minute quantity. The plant hormones are naturally occurring growth regulators that influence the growth and differentiation of plant cells. The growth regulator functions in interaction with the other growth regulators. Such interaction may be synergistic or antagonistic.

Plasmid

These are small, circular DNA present in the bacterial cell besides the main DNA. They can occur independent of the main DNA. Pathogenic bacteria contain these plasmids. During conjugation it is transferred from one cell to another.

Plasmodesmata

These are lateral tubular connections between the cell wall of adjacent cells and help in the transport of solute from one cell to another. It is a connection between two cells and is surrounded by plasma-membrane in the form of tube that is connected between two cells.

Plasmolysis

It is a vital phenomenon that occurs when a cell is placed in a hypertonic solution. Under this condition exosmosis occurs and cell becomes flaccid. With further loss of water the protoplast shrinks away from the cell wall and the protoplasm with plasma membrane lies in the cell centre. The phenomenon of shrinkage of protoplasm is known as plasmolysis.

Plastids

These are present in plant cells and some protists. They are bounded by double membrane. They are of various types depending on the pigment it contains or the storage product present in it. If it has chlorophyll pigment it is known as chloroplast, if it contains other pigment it is called chromoplast and colorless plastid is known as leucoplast while starch storing plastid is amyloplast. Generally all plastids are derived from proplastids.

Plastochron

The time interval between two successive similar events such as appearance of successive leaf primordia at the shoot apex is known as plastochron.

Plastocyanin

It is a copper containing protein that is involved in the transfer of electrons in PSI of photosynthesis. Electrons from PS II are accepted by Cyt b_6/f complex and from this complex to the plastocyanin and from plastocyanin it is accepted by P700 (PSI). It is present in the thylakoid membrane of chloroplast and can freely diffuse along the luminal side of the thylakoid membrane.

Poikilohydric

Poikilohydric term is used for those organisms that depend entirely on water present in the environment. Poikilohydric organisms absorb water directly through their body surface from the environment. They do not have any water conservation mechanism and in a dry environment they remain desiccated and dormant while rehydrates when water is present in the environment, e.g., cyanobacteria, lichens, bryophytes.

Pollution

Undesirable change in physical, chemical and biological characteristics of air, water and soil etc is termed pollution.

Polyamines

They contain two or more amines and are found virtually in all organisms. They are bioactive compounds involved in many processes in plants either singly or in combination. They are polycatronic and thus bind to nucleic acid and plasma membrane. This controls the synthesis or activity of macromolecules and the permeability of membranes. Polyamines are necessary for growth of prokaryotes, eukaryotes and mammalian cells in culture.

Polycarpic Species

These species flower more than one time in their life. After flowering they become vegetative again and before dying they flower at least one more time.

Polysaccharide

Polysaccharides have multiple monosaccharide units and are major carbohydrates involved in both structural and functional aspect of plant life. They usually contain a single sugar but may also contain more than one sugar. The important polylsaccharides in plants include starch and cellulose.

Polysomes

It is used for a strand of mRNA with many ribosomes attached to it. It is found during the process of protein synthesis when translation process occurs. Polysomes are formed so that several peptide chains can be formed simultaneously.

Porometry

It is used for determining stomatal resistance or stomatal conductance.

Pre-Chilling

It is also known as stratification. It is the exposure of a dormant seed to low temperature to induce germination.

Pressure Flow Hypothesis

The hypothesis states that there exists a difference in osmotic potential for the translocation of food downward in phloem. In leaf mesophyll cells, the carbohydrate concentration increases due to photosynthesis. This allows movement of water into it from adjoining cells. These cells then become turgid and pressure develops, which allows movement of food into sieve tube of phloem. The food reaches the root and the root consumes it, therefore a negative pressure develops. This difference in pressure is responsible for flow of food and is known as pressure flow hypothesis.

Primary Body

The tissues developed from the activity of apical meristem are called primary tissues and the body developed from these is called primary body.

Primary Productivity

It is the rate of photosynthesis and storage by producers. In terrestrial ecosystem, producers are mainly herbaceous and woody plants, whereas, in fresh water and marine ecosystems producers are mainly algae. Producers also include chemosynthetic bacteria.

Profundal Zone

The zone occurs below the limnetic zone.

Prosthetic Group

Prosthetic group is the non-proteinaceous cofactor attached to the enzymes. The cofactor is attached more or less permanently to the enzyme.

Protein

Protein is the most important biological molecule that plays an important structural and functional role in plants. The most important role is that it acts as a biological catalyst (enzyme). Protein is formed from amino acids linked through peptide bonds.

P Protein

This is phloem protein. These are strands of proteins found in sieve tubes and sometimes in sieve pores.

Protochlorophyll

Protochlorophyll is the precursor of chlorophyll and its structure is different from chlorophyll because of the presence of double bond between carbon 7 and 8 in ring IV. Chlorophyll is formed by breaking this bond by an enzyme NADPH: protochlorophyll oxidoreductase in the presence of light (in angiosperms) or in dark (gymnosperms and algae).

Protoplasm

It is the liquid portion of cell containing all organelles and chemicals and is bounded by plasma-membrane. The components of protoplasm have important role in the life of a cell.

Protoplast

The protoplasm together with the plasma membrane is known as the protoplast. It is thus a cell excluding its cell wall.

Psychrometry

In fact, vapour pressure of water decreases with decline in water potential. Measurement of water potential using thermocouple is called psychrometry. It can be employed for living tissues and solutions both. Thermocouple psychrometer measures all components of water potential and is very useful in determining water status of plant/soil.

Pulvinus

It is a bulbous structure present at the base of leaf or leaflet. If it is present at the base of petiole it is primary pulvinus, or at the base of pinna then secondary pulvinus. The pulvinus is a motor organ containing large thin walled motor cells. These motor cells respond to changes in turgor and thus drive leaf movement.

Q

Quiescence

When external environmental conditions required for growth of a seed are limiting then the seed is said to be in a state of quiescence.

Quiescent Centre

It is a zone of slowly dividing cells in the centre of root apical meristem just proximal to the root cap. In case of any damage to the root cap or meristem the quiescent zone becomes active and divides to regenerate these parts. The centre has been reported to have the lowest content of DNA, RNA and protein nitrogen in the whole root tip.

Q Enzyme

It is a branching enzyme that catalyses the formation of a branch by transferring a part of chain from one substance to another as in case of starch where some units are transferred from growing starch to either starch itself or other molecule of starch to form α-1,6 linkage.

R

Radial Micellation

It is the radial arrangement of cellulose microfibrils radiating from the centre of stomata. This arrangement of cellulose microfibrils around the circumference of guard cell is responsible for stomatal opening on taking up water. Guard cell expands on taking up water but it does not increase in diameter because of microfibrils which do not increase in length. Thus guard cells increase in length, bend outward to open the stomata.

Radical

It is the embryonal root that gives rise to root in plant.

Reaction Centre

It consists of four to six molecules of chlorophyll *a* and the associated proteins and cofactors. It is the site of primary photochemical redox reaction. At reaction centres light energy is converted into chemical energy. A large number of chlorophyll molecules increase the efficiency in collection and utilization of light energy. In plants two reaction centres are present PS I that has an absorbance maxima at 700 nm and thus designated as P_{700} and PS II (P_{680}) having absorbance maximum at 680 nm.

Reducing Sugars

These sugars have free aldehyde or ketone groups and are thus capable of reducing cupric ion (Cu^{3+}) to cuprous ion (Cu^{2+}) under alkaline conditions e.g., glyceraldehyde having free aldehyde group and dihydroxyacetone having free ketone group.

Reduction

The tendency of gaining electron is called reduction. Any compound that gains electron is reduced while the one that loses electron is oxidized.

Reduction Potential

The potential of accepting electrons is called reduction potential. The compounds that can easily gain electrons have positive reduction potential while the compounds having tendency to lose electrons have negative reduction potential.

Relative Humidity

Humidity refers to the water present in the air. Relative humidity is the ratio of water present in air to the maximum amount of water that the air can hold at that temperature. It can also be defined as the ratio of actual vapour pressure to the saturation vapour pressure. It is generally expressed in percentage as percent relative humidity.

Relative Water Content (RWC)

Leaf water content as a percentage of turgid water content expressed as:

$$\text{Relative Water Content} = [\text{Fresh weight} - \text{Dry weight}/\text{Turgid weight-Dry weight}] \times 100$$

Resins

Found generally in the leaves of conifers. They are mixture of terpenes with 10-30 carbon atoms. The epithelial cells in leaves are responsible for resin formation and they are secreted in the resin duct. Resins protect trees against many kinds of insects.

Respiration

It is a catabolic process in which carbohydrate in the presence of oxygen is broken down into CO_2 and H_2O and energy is released in the process. It occurs in three steps:
i) Glycolysis (in cytoplasm)
ii) Kreb's cycle (in mitochondria)
iii) Electron transport pathway (in mitochondria)

Respiratory Quotient

It is the ratio of CO_2 evolved to O_2 consumed. It depends upon the oxidation state of the substrate. In case of carbohydrate RQ is 1 because 6 molecules of CO_2 and 6 molecules of O_2 are involved. Lipid and protein have RQ less than one because they require more oxygen to complete their oxidation, while for organic acids RQ is more than 1. RQ values are used to identify substrate being used for oxidation in germinating seeds.

Ribonucleases

These are the enzymes that hydrolyze RNA. They act on RNA and release nucleotides whose phosphate group is attached either to carbon 3 or to carbon 5 of ribose unit and N is translocated to other organ.

Ribosome

These are made up of RNA and protein. It is of two types; 70S in eukaryotes and 80S in prokaryotes. It has two subunits 50S and 30S in eukaryotes and 60S and 40S in prokaryotes. Ribosome occurs in the cytosol either freely or attached to the endoplasmic reticulum. The most important function of ribosome is protein synthesis.

Root Gravitropism

Gravitropism is the response of plants to the force of gravity. It may be towards the gravitational pull, i.e., positive gravitropism or away from it, i.e., negative gravitropism. Roots are generally positively gravitropic growing in the direction of earth's gravitational pull. However, it is more accurate for primary root, while tertiary root and higher order roots grow almost horizontally, and this type of growth (both downward and horizontal) helps the root to have a large surface area and therefore facilitates water absorption.

Root Nodule

These are abnormal nodular growth on the root caused by symbiotic nitrogen-fixing bacteria. They are present in the legumes and help in the process of N-fixation. In fact presence of nitrogen in higher amount will block nodule development. Nodule formation occurs when the nod factor produced by bacteria triggers cell division in root.

Root Pressure

It refers to the pressure developed in root when a cell is in turgid condition. This pressure is responsible for upward movement of water in the tracheary elements of the xylem to some extent. This occurs when the pressure is exerted on the liquid content of cortical cells in root forcing the content into xylem vessels to a certain height in the stem. It occurs only when soil has sufficient moisture and humidity is high.

Rubber

This is the longest isoprenoid having 3000 to 6000 units linked in unbranched chain.

RuBP

Ribulose 1, 5- biphosphate is an important substrate that reacts with CO_2 in presence of enzyme rubisco to form a 3-carbon compound 3-phosphoglyceric acid as the first stabler product in the Calvin cycle. It is formed by phosophorylation of ribulose -5-phosphate by ATP and is regenerated at the end of cycle to accept another CO_2 molecule.

Rubisco

It is the most abundant protein on the earth and in leaves. This enzyme catalyses the first step in CO_2 fixation in dark reaction. It reacts with CO_2 to form 3-phosphoglyceric acid in the Calvin cycle and at the last step carbohydrate is synthesized. It acts as both carboxylase and oxygenase. In case of limiting CO_2 and availability of sufficient O_2 it acts as oxygenase resulting in photorespiration. This enzyme works slowly and when light is not limiting it becomes the rate limiting enzyme of the Calvin cycle.

S

Saline Soil

This is such non-acidic soil that contains soluble salt. The presence of salt in such soils affects growth of a plant by interfering with water content of plant causing exosmosis. Salt in soil also affects plant by generating reactive oxygen species.

Salinity

It is the salt content present in dissolved amount in water or in soil.

Saponin

Saponins are chemical compounds found in many plant species. These are formed by the combination of triterpene derivatives with glycoside moieties and are amphipathic in nature. They form soap-like foam in aqueous solution when shaken, e.g., digoxin.

Scarification

It is the process in which seed coat is ruptured to break dormancy. This can be done mechanically either by knife, sand paper or chemicals. Chemicals like alcohol can also be used as they dissolve the impermeable waxy layer on the seed. In nature, scarification occurs by microbial action, digestion of seed coat in the digestive tract of birds and animals and exposure to alternating temperature. Scarification may also occur by fire. These methods used are helpful in seed survival and germination under favorable conditions.

Scutellum

It is present in cotyledons of monocot and is a thin structure that helps to absorb nutrients from endosperm during the period of germination.

Second Law of Thermodynamics

It states that entropy (state of disorder or randomness) of an isolated system which is not in equilibrium will tend to increase and will be maximum at equilibrium. In living system entropy is maximum at the time of death. This law explains irreversibility in nature.

Secondary Body

The tissues formed by lateral meristem are secondary tissues and the body developed from it is called secondary body.

Secondary Phloem

It is formed by the vascular cambium on its outer side during secondary growth and is a complex tissue that can live for several growth cycles. It has sieve elements, parenchyma and even sclerenchyma may be seen.

Secondary Products

These are the end-products of plant metabolism and are secreted from the cell or may accumulate in the cell.

Secondary Xylem

Secondary xylem is formed by cambium during secondary growth. The interfascicular and intrafascicular cambium joins to form a cambium cylinder that gives rise to secondary xylem towards the inner side and secondary phloem towards the outer side by periclinal divisions. Secondary xylem forms the bulk of vascular tissues in woody plants. It consists of tracheary elements, fibres and wood parenchyma arranged in vertical files and parenchymatous cells are oriented at right angle to the longitudinal axis of the plant and they form xylem rays.

Seed Longevity

It refers to the age of a seed or how long a seed can live in its dormant state so that whenever dormancy is over, seed should germinate. Seed longevity has practical utility because seeds are used as genetic resource and as seed gene bank.

Semi-permeable

It is the property of plasma membrane which refers to the selective passage of dissolved substances in and out of the cell. The membrane is only semi-permeable that allows some dissolved substances to pass through it and restrict others. This property is very important for living cells because if the membrane is freely permeable then all soluble substances would leak out making the cell dead.

Senescence

It is the period in life of a living organism that comes after maturity and before death. It refers to the deterioration after development of an organism.

Shaded Leaves

Such leaves are not in direct contact with sun but are shaded by other leaves. These leaves are thinner but larger than sun leaves. They have more chlorophyll (particularly chlorophyll tS) on weight basis and the chloroplasts moves to take a position of maximum light absorption without affecting other chloroplasts.

Shikimic Acid Pathway

This pathway is responsible for synthesis of aromatic amino acids from simple carbohydrate precursors. These precursors may be products of glycolysis or pentose phosphate pathway.

In this pathway shikimic acid is an important intermediate, hence the name of the pathway is shikimic acid pathway.

Short Day Plants

These plants require short days and long nights for flowering. In fact it is the long night that is critical for flowering in these plants, e.g., soybean, sugarcane, tobacco mutant maryland mammoth.

Short-Long-Day Plants

These plants require a short photoperiod followed by a long photoperiod, e.g., wheat and rye.

Sibling Species

Species living in the same region and having almost similar appearance but unable to interbreed due to some incompatibility are referred to as sibling species.

Siderophores

They are secreted by microbes and some higher organisms and act as iron-specific chelates. They may occur as iron-containing or iron-free compounds. Siderophores help in iron absorption by chelation which form soluble iron complex that are taken by active transport. Chemically soil siderophores exist largely as two types; catechols (phenolic compounds with two adjacent hydroxyl groups) produced mainly by bacteria and hydroxamates (peptides) produced mainly by fungi including those of mycorrhizae.

Sieve Tube

Sieve tubes are present in the phloem tissues and are formed when elongated sieve tube elements join other sieve elements. This tube helps in transport of food from leaves. They are living, thick and circular in shape.

Soil Matric Potential

It is a component contributed by matric forces, i.e., the tendency for the matrix to adsorb additional water molecules.

Matric potential takes into account the decrease in the free energy of water, when it exists adsorbed on to the surface of dry soil particles, cell walls and other materials.

Soil Texture

It is soil property used to determine the fraction of sand, silt and clay in soil. It gives the relative proportion of different soil particles in soil.

Solar Constant

It is the power collected per-unit area that is perpendicular to the path of light.

Solar Tracker

It is a device used to concentrate solar light. It concentrates solar reflector or lens towards sun, e.g., Heliostat, a moving mirror that reflects the moving cell to a fixed location. This device is used in solar cell where the concentrated light is directed toward powered device which lies at the focal point of reflector or lens. The leaves of some plants maintain a constant relationship to the direction of sun rays throughout the day. This capability of such plants is known as solar tracking.

Solarization

It is the sterilization of soil and plant material by using sunlight. Sun's radiation is converted to heat by absorption by plant material which gets heated and heat kills the pathogen. For sterilization UV-light is also used which has germicidal property on surface material.

Solvent

It is a liquid that is used to dissolve solid, liquid or gaseous solute. Solvents usually have low boiling point and evaporate easily. A solute dissolved in a solvent forms a solution. A solute can be removed from solvent in a solution by distillation.

Specific Activity

Specific activity is the number of enzyme units per milligram of protein.

Specific Heat

It refers to the amount of heat absorbed per unit mass of the substance to raise the temperature by 1°C. Water has highest specific heat which is responsible for its property of temperature regulation.

Spring Ephemerals

These are short-lived plants that die back to their underground part after short growth and reproductive phase in the spring season.

Standing Crop

The quantity of biological material in a given area in an ecosystem is a measure of standing crop.

Standing State

The quantity of abiotic materials present at any given time in an ecosystem is standing state.

Starch

Starch is a polysaccharide formed from glucose units joined to each other through glycosidic bond. Starch is the major storage carbohydrate which is present as amylose (the unbranched type) and amylopectin (the branched from).

Statolith

Statoliths are present in root cap. These are specialized amyloplast that are involved in perception of gravity. These are also present in the endodermal layers of inflorescence stem.

Sterol

Sterols are amphipathic lipids synthesized from acetyl Co-A by HMG-Co-A reduction pathway. These include campasterol, sitosterol, and stigma sterol and play major role in the physiology of eukaryotes.

Stomata

These are pores found on leaves and are involved in gaseous exchange or loss of water through transpiration. Stomata are present in the epidermal cells of leaves, surrounded by two bean-shaped guard cells that contain chloroplast and guard the opening and closing of stomata. It is through stomata that CO_2 comes out in respiration.

Stomatal Conductance

Stomatal conductance refers to the ease with which water vapour and CO_2 diffuse across the stomata. It is measured in terms of distance travelled per unit time. Stomatal conductance is inversely proportional to stomatal resistance.

Stomatal Resistance

The resistance offered by the stomata for the diffusion of gases, e.g., water vapour and carbon dioxide.

Storage Organ

These are the organs in plant concerned with the storage of energy and water. Energy is stored in the form of carbohydrates. Generally these organs grow underground for protection against herbivores. Such organs are corm, tuber, rhizome, tuberous root, storage taproot, bulb etc.

Stratification

It refers to the low temperature treatment that is used to break dormancy.

Stress Proteins

Stress proteins are certain proteins induced and synthesized in response to external stress (heat/osmotic/drought). The proteins formed may vary with the type of tissue, growth conditions and plant species.

Stroma

It is a thick fluid in which the dark reaction of photosynthesis occurs. Calvin cycle occurs in the stroma utilizing the products of light reaction to synthesize carbohydrates.

Suberin

It is a hydrophobic material that prevents entry of water into tissues. In root the casparian strip is made up of suberin which allows only symplastic movement of water across the endodermis. Suberin is also found in outermost layer of bark preventing loss of water from tissues that lie below it. These are a complex mixture of long chain fatty acids and alcohols and as in cutin, there may be fairly large percentage of aromatic compounds.

Succulent Plants

These plants have water-storing capacity and are thus adapted to dry climate. In case of high temperature and low precipitation these plants store water to survive the dry period. Such plants may store water in leaves, stems or roots and appear swollen after storing water. Succulent plants open their stomata only at night to minimize water loss through transpiration during daytime. They have reduced leaf surface, fewer stomata, waxy, hairy or spiny outer surface and stem as the main site for photosynthesis.

Sulfate Reduction

Sulfur is taken up by plants in the form of sulfate but it has to be reduced for further processes. The reduction occurs in the chloroplast in the shoot where sulfate is reduced to sulfide and is then incorporated into the amino acid cysteine which is an important sulfur donor for most sulfur containing organic compounds like proteins, thiols or sulfolipids.

Sulfoxides

It is an oxidized sulfide and has sulfinyl as the functional group attached to two carbon atoms.

Symplast

It is the inner side of the plasma-membrane in which water can freely diffuse. It allows direct movement from cytoplasm to cytoplasm. Its pathway is from epidermis to cortex and then to endodermis. From endodermis it goes to pericycle and then to xylem for transport.

Symport

An integral membrane protein is involved in the transport of two substances in the same direction across the membrane. Such transport involves the facilitated diffusion of one substance that is coupled with the active transport (against concentration gradient) of the other substance.

T

Terpenoids

These are lipids present in living organisms and are formed from isoprene units joined together and modified in various ways. They are also known as isoprenoids. They play important role in plants and are used because of their aromatic quality. They have pharmaceutical application and are responsible for smell, taste and colour of various plants, e.g., scent of *Eucalyptus* or flavor of cinnamon and cloves. They may be added to proteins to enhance their attachment to cell membrane.

Thermal Radiation

Any object that is hot and emits light from its surface known as thermal radiation. It is an electromagnetic radiation. The hotter the object more is the light it emits. Thermal radiation may be defined as the light emitted from a glowing incandescent bulb.

Thermodynamics

It is a branch of science that deals with the effects of changes in temperature, pressure and volume in any system by observing the motion of particles it also tells how the system responds to these changes in the surroundings.

Thermonasty

It is a nastic movement that depends upon temperature changes. In certain flowers temperature affects their opening.

Thermoperiodism

It is the response to difference in temperature at day and night. A cycle of temperature works with the light-dark cycle and influences plant growth, primarily for floral production. A plant can grow vigorously if day temperature is high alternating with low night temperature and the reverse would discourage growth.

Thigmomorphogenesis

It is the change in growth pattern of a plant in response to mechanical stimulus of touch.

Thigmonasty

It is a nastic response of plant to stimulus like touch and heat and is independent of the direction of the stimulus. Thigmonasty is observed in venus fly-trap in which the trap closes on the visit of an insect. Turgor pressure and action potential are responsible for this phenomenon. In *Mimosa pudica* the leaves droop on touch and this is also due to action potential that reaches the pulvinus. Thigmonasty occurs at a very high speed.

Thigmotropism

It is the response of plant to stimulus and occurs in the direction of the stimulus. For example, the root growth follows thigmotropism because it grows away from rock in soil.

Thioredoxin

This is oxido-reductase enzyme that catalyses the reduction of proteins by cysteine thiol-disulfide exchange and thus acts as antioxidant. They have a thiol-disulfide active site.

They are important for the life of plants. They act as electron donor to peroxidases, ribonucleotide reductase and also in the electron transport in photosynthesis.

Third Law of Thermodynamics

It states that at an absolute zero temperature entropy tends to be minimum and is known as absolute entropy. The temperature of a system increases its entropy increases.

Threshold

It is the minimum limit to any quantity or stimulus that is needed to induce a response.

Thylakoid Membrane

It is the membrane surrounding the thylakoids separating one thylakoid from other in the grana stack. It is the site of light reaction of photosynthesis. Structurally it is similar to the plasma membrane.

Tonoplast

It is the membrane surrounding the vacuole in plant cells and has structure similar to plasma membrane. It separates protoplasm from vacuole sap.

Torus

It is the swollen structure that is present in the centre of bordered pit and it is formed only of primary wall material. The function of torus is to act as a valve blocking the pit when the pressure on one side exceeds pressure on the other side. Torus is found in trachieds and vessels of many gymnosperms.

Totipotency

It is the ability of a differentiated cell to regenerate into a new plant. The differentiated cell forms an undifferentiated callus that grows into a new plant. This property of some differentiated cells shows that cells retain their genetic information even on being specialized or differentiated but all the information are not used up by the cell at any given time.

Trachieds

These are non-living cells of the xylem and originally form a single cell. They are present in all vascular plants. Tracheids have perforation on their common wall in the form of pit pairs. Together with vessels they help in water conduction.

Transamination

This reaction involves the transfer of an amino group. Either an amino group is transferred to α-ketoacid reversing the α-ketoacid to amino acid and amino group to α-keto acid or alkyl amino de-amination may occur in which an amine or amide ion is substituted by another amine or amide. In human body transamination occurs to form non-essential amino acids.

Transduction

It is the process of transfer of genetic material from one bacterium to another through the agency of a temperate bacteriophage. The phage when released from bacteria takes some of its bacterial gene and on attacking an other bacteria it inserts its bacterial gene into it. This incorporation of different genes in one organism forms a transformed bacteria and the process is known as transduction.

Translocation

It is the transfer of a substance from one location to other. Food synthesized in leaves is transported to other parts through phloem. Phytohormones are translocated from site of origin to the site of action. Water and inorganic salts move from roots upwards through xylem.

Transpiration

It is a vital process in which loss of water in the form of water vapors occurs from aerial parts of the plant. Water loss through epidermal cell of leaves and herbaceous stem is cuticular transpiration, while if water loss occurs through lenticels it is lenticular transpiration whereas, major part of water is lost through stomata, which is known as stomatal transpiration. It generally occurs during day as during daytime stomata are open. Transpiration has many important functions. Besides keeping the plant cool it helps in the ascent of sap.

Transpiration Efficiency

Transpiration efficiency mainly expresses water use efficiency at the single leaf level. Stomatal structure is supposed to maintain access to atmospheric carbon dioxide while controlling diffusion of water vapour.

Transpiration Stream

The flow of water from the soil through the xylem resulting due to transpiration is transpiration stream. It may help mineral absorption from the soil and transport to the plant.

Triacontanol

It is a fatty alcohol that is present in the cuticle waxes of plants and acts as a stimulant for growth.

Triglycerides

These are esters of fatty acids with glycerol. One molecule of glycerol is attached through an ester bond with three molecules of fatty acids at their hydroxyl group. It is in this form that fat exists in food and in body.

Triple Response

Ethylene is a gaseous plant hormone that affects plant growth by its triple response. It is so called because ethylene application results in three different responses on the same plant for one physiological effect, i.e., growth (i) it causes reduction in elongation (ii) swelling of hypocotyls that increases lateral growth and (iii) change in the direction of growth from longitudinal to radial.

Tropic Movement

This movement occurs in response to an external stimulus. The stimulus affects the growth movement of a plant or plant parts either in the direction of stimulus itself or away from it. This directional growth movement takes place as in the presence of stimulus unequal growth occurs on the two sides of an organ. This may also be attributed to increased auxin levels in the tissue.

Turgidity

It develops in a cell when it absorbs water. Water enters into cell through osmosis making the cell turgid and as a result pressure is developed inside the cell. Turgidity of cell is important for its normal growth. Loss of turgidity leads to wilting of leaves and drooping of shoots.

Turgor Pressure

This pressure develops in turgid cells. When a cell is placed in hypotonic solution water enters into the cell and makes it turgid. In such cells a hydrostatic pressure is exerted on the cell wall called the turgor pressure. This pressure helps in cell enlargement and keeps leaves erect and expanded.

Turnover Number

Turnover number of an enzyme is the number of substrate molecules transformed per minute time by a single enzyme molecule.

Tyloses

These are balloon-like ingrowths protruding into the tracheary elements of older wood. Tyloses are filled with tannins, resins or other pigments of commercial use. Tannins and resins present in the tyloses provide strength to the wood.

Tyndall Effect

It is the property of colloidal solutions in which light is scattered by small colloidal suspension and makes the path of light visible. Although these colloidal suspensions are not visible to naked eye but on scattering light they make the path of light visible.

U

Unit Enzyme

One unit of an enzyme is defined as that amount causing transformation of 1μmol of substrate per minute at 25°C under optimal conditions of measurements.

Ureides

These are compounds containing urea and are formed in root nodules during nitrogen fixation. They are also formed during metabolism of purine bases. On breakdown ureides form urea that accumulate in toxic amount in Ni-deficient plants. Since ureides are formed in all plants that show the necessity of urease and nickel for plants.

Uttradian

These are the rhythms in metabolic activity that have period less than 24 hours. The period can even be measured in minutes.

V

VAM

It is a type of mycorrhizal association in which fungi show symbiotic association with plants. VAM fungi penetrate the root cell forming vesicles and arbuscles and its hyphae are spread in the soil. Thus they provide a connection between soil and plant helping in transport of minerals to plants and carbon compounds into soil. VAM also helps in growth of a plant. An important function of VAM is to provide resistance to environment stresses.

Vacuole

It is an organelle present in the cytoplasm of a cell. It has a fluid called vacuolar sap containing organic and inorganic compounds. Vacuole is surrounded by vacuolar membrane called the tonoplast. The vacuole accumulates reserve food material, metabolic byproducts and also various toxic compounds in tolerant plants.

Vapour Density

It is defined as the mass of a given volume of gas divided by mass of same volume of hydrogen gas at the same temperature and pressure. It may also be given as half of the molar mass of that gas whose vapour density is to be known.

Vapour Pressure Deficit

Reduction in the partial pressure of water vapour in air compared to that of leaf of a plant measured in terms of pascal or bars. The difference in vapour pressure acts as a driving force for transpiration.

Vernalin

It is a stimulus that is transmitted from apical meristem to other parts to induce flowering. Vernalin works in association with florigen, the flower-inducing stimulus.

Vernalization

It refers to the cold treatment given to plants to induce flowering. The temperature that is effective for vernalization lies in the range of 0-5°C. Vernalization does not induce flowering but it only prepares the plant to flower. It stimulates the shoot apical meristem, which translocates the stimulus to other plant parts.

Vialaxanthin

It is a natural xanthophyll pigment, derivative of α-carotene. It is orange in colour.

Viscosity

It is defined as the resistance offered to the flow of liquid. A non-viscous fluid will spread out. Viscosity is internal resistance to flow and is a measure of friction present in the fluid.

Vitamins

These are essential compounds needed for the proper growth and functioning of an organism. They are needed in small amount and not synthesized in the human body. Heterotrophs depend upon plant and microbes for vitamins. Vitamins are of two types: fat soluble- A, D, E and K and water soluble- B and C.

W

Warburg Effect

It refers to a condition of excess oxygen which inhibits photosynthesis by inhibiting CO_2 fixation. In the presence of high oxygen concentration rubisco acts as oxygenase and not as carboxylase resulting into decreased CO_2 fixation.

Waterlogging

Water is very essential for plant growth and survival but the requirement of water for plants differs from one plant to another. For a plant that needs less water, waterlogging will cause adverse effects on root system, leaves and on the whole plant. It develops a stressful situation like water stress.

Water Potential

It is defined in terms of free energy of water compared with free energy of pure water in any system at the same temperature and pressure. Water potential of any substance decreases on addition of solute.

Water Stress

It occurs when the soil has insufficient moisture and transpiration is high. Due to water stress plant growth is affected. In extreme stress condition plant may wilt and die.

Water Use Efficiency

Water use efficiency is the dry matter produced per unit of water used in evapotranspiration. At the single leaf level it is the ratio of carbon assimilated to water lost in transpiration.

Waxes

These are long chain hydrocarbons with little or no oxygen. They may also be long chain fatty acids esterified to long chain alcohols.

X

Xanthophylls

The xanthophylls are oxygenated derivatives of the carotenoids. Xanthophylls are yellow coloured pigments present in the leaves of plants in plastids. Carotene is oxidized forming xanthophylls which have hydroxyl group and are more polar in nature compared to carotene. They function during photosynthesis as accessory pigments.

Xerophytes

These are plants that can survive in water deficient conditions. In soil having low moisture or in such environment where evapotranspiration is higher and precipitation is lower only xerophytes can survive. These plants are generally found in desert habitat but may be found in moist habitat also. Xerophytes exhibit adaptations to survive under drought conditions. They have sunken stomata to reduce transpiration, deep roots to absorb water from underground sources. In these plants water is stored in organs like stem, leaf and root tissues.

Xylem

It is a complex tissue that helps in transport of water and minerals. It also provides mechanical support. Xylem has both living and non-living cells known as xylem elements. The non-living elements are (i) tracheary elements, i.e., tracheids and vessels which are involved in water conduction (ii) xylem fibres which provide mechanical support. The living element is xylem parenchyma which is concerned with storage and translocation of ergastic substances.

Xylem Parenchyma

It is present in the primary and secondary xylem and is a living cell that stores food material and also helps in the lateral conduction of water and minerals.

Z

Z Scheme

In the process of electron transport during photophosphorylation in photosynthesis, link between photosystem I and photosystem II is referred as the Z-scheme. Electrons move in Z-scheme from water to $NADP^+$. On being excited by light energy, PS II (P_{680}) is changed to its excited form and is photooxidized releasing an electron that is accepted by pheophytin. From pheophytin it is accepted by QA then by plastoquinone. The oxidized PS II is reduced by the electrons coming from the splitting of water. Water is oxidized by the positive manganese ion that is associated with thylakoid membrane and pass electron to the PS II. From PS II the electron enters the PS I complex through Cyt b/f complex and plastocyanin. PS I (P_{700}) is excited by light that passes electron to $NADP^+$ through ferridoxin forming NADPH.

β – Carotene

It is a reddish-orange color pigment found in many plants and fruits. It is the main source of vitamin A being its precursor. It is found in yellow and orange fruits like mango, papaya and in green leafy vegetable like spinach. They also act as accessory pigments and protect the chlorophyll from photo-oxidative damage.

β- Oxidation

It occurs in mitochondria and/or peroxysomes and in the process fatty acid is broken down to acetyl Co-A which enters Kreb's cycle.